William Day

The Horse

How to Breed and Rear Him

William Day

The Horse
How to Breed and Rear Him

ISBN/EAN: 9783744666794

Printed in Europe, USA, Canada, Australia, Japan

Cover: Foto ©berggeist007 / pixelio.de

More available books at **www.hansebooks.com**

THE HORSE

HOW TO BREED AND REAR HIM

RACE-HORSE
HUNTER
HACK
TROOP-HORSE

DRAUGHT-HORSE
SHIRE-HORSE
CARRIAGE-HORSE
PONY

BY
WILLIAM DAY

AUTHOR OF
'THE RACE-HORSE IN TRAINING,' 'REMINISCENCES OF THE TURF,' ETC.

SECOND EDITION

LONDON
RICHARD BENTLEY AND SON
Publishers in Ordinary to Her Majesty the Queen
1890

[All rights reserved]

PREFACE TO FIRST EDITION.

I HAVE taken it on myself to write the following work in the belief that such a book is really needed at the present time. Perhaps I have also been persuaded to the task by the kind reception accorded to my former volumes. I had something to say of a practical kind, such as only a breeder and trainer of many years' experience could say; and I have done my best to put the results of that experience into these pages. My treatise on the horse, therefore, should contain much in the way of advice that may—and will, I hope—prove of value and assistance to others. There cannot be many such books recently published, for I do not know of one which is the work of a truly practical and experienced man.

I am quite prepared to find exception taken to some of my views; for I am well aware that, on several points, I hold opinions of a pronounced kind. It should be remembered, however, that I have formed those opinions after a lifelong acquaintance of the most intimate sort with the subjects of which I treat.

I have a good deal to say on the improvement of our thoroughbred stock, undoubtedly the best in the world. My views are fully set forth in the following pages, and it is only necessary to repeat here my firm conviction of the truth of all I have said, and the crying need of measures such as those I have advocated. Nor have I confined myself to the thoroughbred, but have also discussed many things relative to other kinds of horses.

To one topic in particular I desire to draw the widest attention. It is that entered upon in the chapter entitled 'The Half-Bred on the Farm,' and discussed at length in that and some following chapters. I have, in these, endeavoured to point out the relations which the horse bears to the social and political economy of the nation; and I believe I have demonstrated means by which, through and by the horse, immense benefit may accrue to the agricultural interest throughout the United Kingdom of Great Britain and Ireland.

Some delay has occurred in the production of this book, due to circumstances over which I could exercise no personal control. I mention this merely to account for a few dates scattered through my pages, which might, without this explanation, be regarded as not sufficiently recent; such as concern Appendix II., for example.

For the rest, I am abundantly conscious of many defects and imperfections in my work. But I leave it, such as it is, to the judgment, and, I hope, the favour, of the readers and critics who have so kindly received my former efforts. And I take this opportunity of thanking both, but especially the latter, for their previous indulgence.

W. D.

SALISBURY,
    September, 1888.

CONTENTS.

CHAPTER I.

INTRODUCTION.

PAGE

My aim in writing—Need of special qualifications—The question of improving the breed of our horses a national question of the hour—My experience as a breeder—List of Alvediston breeding stud—Absence of authoritative modern teaching—Excellence of theory, but need of practical knowledge—My reliance for correctness on experience alone—Existing works not entirely satisfactory—Cecil's 'Stud Farm'—'Stonehenge.'

Needful care of minor details—Value of perseverance—The method and order of the work set forth in detail . . . . 1—11

CHAPTER II.

BREEDING AS A NATIONAL QUESTION.

Absence of works on breeding horses of useful kinds—Number and value of horses other than thoroughbreds—One class taken as an example—The farmer's interest in breeding the hunter.

Hunting in old times—Hunting veterans: Mr. Radclyffe, Lord Radnor, Mr. Trelawny, and others—Stag, fox, and hare hunting—The Buckhounds—Lord Pembroke's harriers—The Prince of Wales in the field—Want of good horses.

Wide discussion of possible improvements in breeding—Opinions of Mr. Craven and Mr. Gilbey—Government views—Testimony to prevalence of injudicious selection, and to certain benefits from greater care—Our present deficiencies—The show at Newcastle: unsound stallions—Dearth of good horses in the West—Our natural advantages—No difficulty to increase numbers—The need is 'quality'—To be obtained by selection, and how—Practical, not scientific, information wanted . . . 12—25

CHAPTER III.

HISTORY OF THE HORSE.

Numerous authorities on the subject—Prehistoric remains—Zoological classification—General description—Always much valued—Original habitat and gradual extension—Early use in Egypt—Biblical testimony—Claim of Arabia as its origin examined—My own negative opinion confirmed—Use in Persia and Palestine.

The horse in Britain—Records of large animals in a wild state—Used in the invasion by Cæsar—Cæsar's testimony to its previous subjugation in Britain—Already an animal of value—Of necessity of mixed breed—First trace of distinct crosses . . . . . . 26—39

CHAPTER IV.

PROGRESS OF RACING AND ITS RESULTS.

Decline of cruelty in sport—Corresponding increase in racing—Earliest records—Homer—Racing colts in Greece.

Racing in England—In Athelstan's time—Henry II.—Henry VIII. and Elizabeth—Encouragement given to breeding by the Stuarts—James I. imports horses—The *Markham Arabian*—Cromwell's *White Turk*—Charles II. a genuine benefactor—The royal mares—Commencement of the 'Stud-Book'—Racing at Newmarket—Queen Anne—Importation of noted Eastern sires—George IV.—Anecdote of the 'Sailor King'—H.R.H. the Prince of Wales as a breeder and owner—The Hampton Court Stud—Foreign triumphs in breeding only exceptional . . . 40—50

CHAPTER V.

ORIGIN OF EXISTING BREEDS.

Sources of information—Earliest crosses—English horse valued in Saxon era—Spanish strain—Horses in the 'Stud-Book' in 1793—The *Byerly* and *Darley* Arabians—*Curwen's Bay Barb* and its produce—*Flying Childers'* fabled speed—The *Godolphin Arabian*; no pedigree—Description —His first produce—Story of the black cat—Result of the crosses; three famous sires: *Matchem, Herod*, and *Eclipse*—Performances of the latter—*Marske*—*Highflyer* —The successors of the three in our own day . 51—62

CHAPTER VI.

REAL VALUE OF THE ARAB STRAIN.

Alleged superiority of the Arab—Upton's 'Newmarket and Arabia'—Advantage of the cross apparent only in later generations—The *Godolphin Arabian* as a sire; not equal to present stallions—Early Arabs not successful on the course or at the stud—Instances of improvement as the cross recedes: the *Bald Galloway;* 18 stone to victory; other examples—Instances to the contrary from earlier horses—Failure of the cross since *Eclipse's* time—Authorities on my side: Buffon, Goldsmith, 'Nimrod,' Admiral Rous, Cecil, the *Sporting Life*—Covering fees of noted Arabs —A contrast—How much 'temper' may be due to the strain —Protest against revival of the practice—Failure of Hampton Court experiments . . . . 63—79

CHAPTER VII.

IMPROVEMENT OF THE MODERN HORSE.

Existing disbelief—One cause of this: 'short courses'— Difficulties of the breeders of judging—'Gameness' of our present horses—Admiral Rous's testimony—What horses really did a hundred years ago—'Sweeping the board' in those days; a ten-guinea prize—Performances of *Eclipse* examined—*Eclipse* and *Touchstone* contrasted on the turf and at the stud—The lesson therefrom—Advantageous effects of climate; the Arabian 'a curious exception'— Our improved fat and lean stock—My Russian pig; 'Early Porcine Type'—Spread and appreciation of the English horse in other countries; 'Stonehenge' thereon; what 'Weatherby's List' says; 'Nimrod's' testimony— Progress in the colonies—New Zealand. . . 80—94

CHAPTER VIII.

CLIMATE AND OTHER INFLUENCES.

Climate and soil must be suitable—Abundance of eligible farms—Effect of our climate on horses in size and shape—Undesirable localities—Goldsmith on climatic effect on dogs; on improvement in cattle—The wild herd at Chillingham—Progressive advance in breeding—Increased weight and value of sheep—Weight of a modern ox—Contrast of past and present times.

Other influences—Further requirement in selecting farm—Abundant pasturage—Sufficient room—Effect of exercise on action—Essential need of it—Example from greyhounds . . . . 95—105

CHAPTER IX.

SLIPPING AND STERILITY.

Wide extent of loss from this cause—How ewes are treated—My experience shows that it is a disease—Early exhibited at the stud—Twins also a loss—Inherited and accidental causes.

Nature of malady examined—Examples through three generations—Dead foals lead to barrenness; examples—Afflicted mares die early—Slipping leads to barrenness; examples—Extent of malady shown in 'Stud-Book'—Examples from one hundred representative cases—Must end in loss.

Accidental causes—Care needed during pregnancy—

Removal of refuse in paddock and yard—Fright—Bad smells—Must separate barren mares—Extraordinary instance at Alvediston from eating fallen leaves.

Result of examination — Slipping and sterility go together—Exemption of Irish mares suggests the question of climate and treatment . . . . . 106—118

CHAPTER X.
TEMPER, SIGHT, ROARING, SPLINT.

Temper of horses important—Traced through generations—*Young Trumpeter* and his offspring; *Little Tom* and *Bugler*, *Sigmaphone*—Narrow escape of *Crucifix;* practical loss of her second race; always started badly; evil recurrent in *Surplice*—Examples of complete failures; *Entre Nous* and *Vittoria*—Defects due to temper—Danger in the paddock—Effect of improper punishment—Influence of breaking on temper; an opinion on the Sledmere process and my own opinion—Mr. Farquharson's ill-success.

Faint-heart a source of great loss—Examples in *Allie Slade* and others.

Defective sight transmitted and accidental—*Drogheda* and *Defence*—Cause of accident and expense; instance with Mr. Sadler—Inherited nature of disease proved by decrease in cataract.

Roaring a transmitted disease; often dormant; instance in *Ormonde*—Supposed remedies—Influence of climate; *Belladrum* in South Africa—Lessened by better treatment.

Splints, if serious, may be inherited—Need to discern between accidental and inherited maladies . . 119—135

CHAPTER XI.
CONSANGUINITY, CROSSING, COLOUR.

Difference of opinion on in-and-in breeding—Purity of blood essential; value of crossing for the purpose—Tendency of domesticated animals run wild to breed back—Early neglect of the horse—Darwin and Youatt on effect of selection in breeding—Darwin on crossing and fertility—Effect on the human race—The 'daft' villager—Loss of gameness in fowls—Greyhounds improved by cross with bull-dog—Effect on the race-horse—*Melbourne's* produce—The thoroughbred cross in all our horses.

Transmission of colour through a century—Marked instance in *Crab* and his offspring—Descent of *Chanticleer* stock—*Grey Friar*—*Birdcatcher's* grey flanks—Colour not material, but serves as proof of transmitted qualities from ancestors . . . . 136—148

CHAPTER XII.

GENERAL PRINCIPLES OF SELECTION.

Books not much help; experience and observation the true guides — Transmission of peculiarities; unsoundness; lameness—'Stonehenge' on the subject—Pritchard's views—Control of man over hereditary modifications in animals; Darwin—Confirmation; necessity for matching sire and dam in this respect—Sir Tatton Sykes' stud, and my opinion of it.

Judgment of powers not possible without trial; instances in point—Pedigree a safeguard in selection—The breeder's difficulties—Stallion and mare must possess like features—Speed and stamina desirable on both sides—Diversity of opinion on in-and-out crossing—Why so few mares breed good horses; a theory as to the reason; examples.

The best age to put mares to the stud—Lord George Bentinck's experiments—Various illustrative instances—The rule, and exceptions to it—Early breaking-in not a deterrent to long work; management; hunters have lasted long—Work of English race-horses compared with that of American trotters.

Summary of the principles of selection . . 149—169

CHAPTER XIII.

SELECTION OF THE MARE.

Selection involves complex study—The details to be considered—Examples drawn from thoroughbreds are applicable in the case of any other class.

Breeding—Choose either distinguished or young mares—Relationship to good horses my great point—Mares of good blood justify their selection on that ground; examples—Mr. Wreford's experiences—A good performer, if of bad stock, is generally disappointing—Some 'good-looking failures'; no reason assignable—The element of chance—Three lucky purchases: the *Cervantes* mare, *Grace Darling, Octaviana*—Three unfortunate purchases: *Marie Stuart, Mandragora, Agility*—The lessons these cases teach us.

Shape and size—My preference—Small mares of strong frame the best—*Glee;* other examples.

Action—Mares with bad action often failures—Mares with good action produce clever horses—Examples—This point should be studied—Action may be developed, but originates from breeding—Curious instance in a dog.

Constitution—Soundness and absence of all disease essential—How to judge of this—*Alice Hawthorn* and *Beeswing*—The pedigree a test.

Speed and stamina—Fast mares preferable for the stud—Summary of all the points to be considered . 170—186

CHAPTER XIV.

SELECTION OF THE STALLION.

Health a first essential—Difference in cases of mare and stallion—Power of the stallion to transmit diseases; case from *The Veterinary Journal*—The best horses, of pure strain and free from defects, should alone be chosen—We keep too many inferior animals.

Breeding a point of great importance—Pedigree must be closely scanned—Temperament and capabilities of ancestors to be studied for several generations.

Size and shape—Middling-sized horses most successful—*Eclipse, Touchstone*, and many other examples—Failures of large horses; *Prince Charlie* and others—*Stockwell* unique in this respect—Result of comparison—Shape, another essential—Only one good form—Description of my ideal.

Effect of using inferior stallions—Subsequent offspring by other horses affected—Cecil's opinion—One cross sullies a mare for life—Overwhelming evidence in proof of this—Sir Gore Ouseley's cross between a mare and zebra; result—Chestnut mare and quagga; result—Darwin on the subject—Necessity of precaution therefore . . . 187—198

CHAPTER XV.

THE THOROUGHBRED MARE.

Respective influence of sire and dam on their offspring; discussion—Instances—Suitability essential.

Opinions as to respective ages of sire and dam—Some notable first foals—Youth evidently no drawback on either side.

Inequalities of size considered—Large animals have poor progeny—Remedying of defects by suitable selection—Different methods employed by breeders—Opinion of an expert—Experiment of the late Earl of Derby—Size should be improved by gradual selection.

Curious fact that mares breed best from particular stallions—Cases in point—Breeding a winner proves the suitability of the cross—*Defenceless, Flying Duchess*, and *Elcho's* dam—Innumerable proofs of my contention.

The question whether young mares are better than those which have raced long, or the reverse—My view that we must always look to pedigree—Three examples on each side—Solution of the problem . . . . 199—212

CHAPTER XVI.

THE THOROUGHBRED STALLION.

Former partiality for the strains of *Highflyer, Eclipse,* and *Matchem*—Present liking for those of *Touchstone, Voltigeur,* and *Stockwell*—The latter the best horses yet seen—Their pedigree and descendants.

Evils of breeding from your own stallion; exception

and proof—Rank fallacy of the practice demonstrated—
The experience of various gentlemen—Numerous instances
—The unavoidable deduction therefrom.

Age of the stallion—Capriciousness of evidence—No
rule can be fixed upon—Various opinions.

The Queen's Plates—Prohibition of geldings—I demur
to this—Many inferior stallions should be cut—Encouragement to geldings might improve our breed—Mr.
Craven's opinion—The matter gravely considered—The
knife necessary.

Anomalies in breeding—Trainers' systems—Public
running a farce—A typical trainer—His failure at long
distances — Speed *versus* stamina — The historian's
duty . . . . 213—226

CHAPTER XVII.

MATING.

Meeting of the sexes—Time to put mares to the horse—The
mare should be hobbled when covered; reasons.

The moment of conception no concern of the breeder's
—Repetitions of service to be avoided—Mr. Sadler's ex-
experience in this respect—A case that occurred to me.

Virgin mares to be cautiously dealt with—Attention
necessary—Bringing them to the stallion—Care to be
taken not to exhaust the horse's powers.

Thoroughbred mares—Consideration of time of mating
with regard to racing rules—The fourteenth of February
—Period of gestation of mares—A popular theory—
Service of barren mares—Likelihood of procuring pregnancy—Modes of causing the mare to be stinted.

Two systems of covering—My view—Instance in the
case of bitches—An expert's opinion.

Treatment of mare after service—Turning—Frequency
of covering; when desirable and not—Indications of being
at use—Opportunity to be seized—Disadvantages of late-
born horses—A year's rest for brood-mares; my opinion
on that head. . . . . 227—237

CHAPTER XVIII.

TREATMENT OF MARE, FOAL, AND YEARLING.

PAGE

An interesting study—Thorough knowledge of the subject necessary—Causes rendering mares unfit for reproduction—Turning them out too suddenly a chief one—The best system to follow.

Treatment after service and during gestation; of young mares; of mares with foals—A common cause of abortion—The same in sheep—Feeding and housing in summer and winter—Handling the foal—Attention required—Benefit of moisture to the development of the feet—Excessive moisture detrimental—Good feeding essential.

Time of foaling—Previous indications—Delivery—Abnormal presentations—The after-birth; its retention a peril—The new-born foal—Immediate treatment required by it and by the mare—Food—Suppository.

The suckling foal—How to ensure its proper development—Giving it bruised corn—Dieting the mare—The foal's mid-day meal—Time for weaning—Separation of dam and foal—Variety of food—Articles to be avoided—Condiments a mistake—Physic or no physic—My opinion and Mr. Robinson's.

The yearling—Free exercise imperatively necessary—General principles of treatment—The feet, and how to manage them—Irregular growth to be checked—Paring and rasping—Proper development of the legs depends on careful attention to the growing hoof . . 238—252

CHAPTER XIX.

TREATMENT OF THE STALLION.

Customary neglect of certain points—Common mode of treating the stallion—Pampering; want of exercise; neglect of the feet—Evil effects—How the horse ought to be kept—Exercise—Comparison with brood-mares.

Disease of the feet—Examples—My experience; what

I argue from it—Rates of mortality—Causes of death—
Primary reasons for such effects—Negligent stud-grooms
—Treatment of the country stallion compared with that of
the thoroughbred—Comparison of results—Mr. Robinson's
bull—His principle.

The bowels—Constipation and diarrhœa—How to meet
them—Prevention better than cure—Stallions at first
indifferent to the mare; examples—This fault overcome
by perseverance . . . . 253—262

CHAPTER XX.

THE HUNTER AND TROOP-HORSE.

Common origin of all kinds of horses—Hunters, hounds,
and foxes of the past—Superiority of English troop-horses
—Abundance of hunting sires—Thoroughbreds in North
Devon; an article in *The Field*.

Employment of thoroughbred stallions for getting
hunters—Modern hunters of good blood—Common-bred
hunters too slow—Hounds faster than any horse.

Selection of mare and stallion—The thoroughbred
cross—Shape and size—Summary of the points to be
studied—Mares hunted too long; a more judicious plan
—Early breeding—Early working—How the foals should
be treated.

Weight-carriers—Special selection of sire and dam
requisite—The Grand National Steeplechase, past and present—Nearly thoroughbred hunters the best—Anecdote
of Mr. T. Ashton Smith—Inexpediency of putting hunters
in harness . . . . 263—274

CHAPTER XXI.

THE HALF-BRED ON THE FARM.

Breeding half-bred horses remunerative for farmers—Mares
can be worked as well as bred from—The preferable sort
—Prices and profits—Depreciation in values of cattle,
etc.—Breeding good horses a boon to farmers.

A calculation—Twenty-four mares; their produce and profit—Working and breeding—The system tried and proved successful—Additional profits.

Comparison of the ordinary shire-horse with the thoroughbred and half-bred—Respective pace—Time of ploughing an acre—Distance travelled—Gain in time, and in amount of work accomplished—Calculation of the saving and profit on a large farm.

The two sources of profit; total—Amazing result—Objections answered—Lord Lonsdale's experience; Mr. Robinson's; my own—Employment of a stable-keeper; why—Saving thereby—Have shire-horses deteriorated? —Comparison with coach-horses—Arguments—The answer. . . . . 275—289

CHAPTER XXII.

THE HALF-BRED v. THE SHIRE-HORSE.

Continuation of the argument—How the men would be affected—Working hours—Employers' benefits—Piecework instead of time-work—Mr. Radcliffe and Mr. Sabin —Opinion of the latter—Differences in labourers' day; work done always the same—Extraordinary apathy of farmers—Time taken to plough an acre; instance—The eastern counties and the southern.

Waste of time—A calculation—Present system induces men to idle—My plan will stimulate them—Return of the 'good old times' for agriculturists.

Faster speed on the road also a source of profit—Calculations and comparisons—Ignorance of farmers concerning the capabilities of their horses; this exemplified—An anecdote—American surprise at our slow cart-horses—American competition; causes of its success, and the remedy—Farmers' notions as to time and work—The distance travelled in ploughing an acre worked out—Deductions from the foregoing . . . . . 290—311

CHAPTER XXIII.

COMPARISON OF DRAUGHT-HORSES.

Other draught-horses *versus* the cart-horse—The coach-horse —Rates of speed—Fast and slow—The omnibus-horse; loads and distances—The post-horse—The van-horse; work performed—Good comparison—A revelation to farmers.

Dr. Johnson's ideas of progress—My reply thereto; pertinent to the question—Improvements everywhere; our agriculturists alone unprogressive — Labour and labourers—Systems in fault—Piece-work encourages industry—An example from Mexico—Faults of the British farm-labourer.

Apathy of our farmers—Advice given in vain—The secret of depression—Agricultural statistics—Amazing expenditure—Immense saving possible—The means— Utilization of waste lands—Production—Comparative table of draught-horses and their several capabilities— Sorry figure made by the shire-horse — The largest and strongest draught-horse proved to be the most inefficient . . . . 312—325

CHAPTER XXIV.

THE CONSEQUENCE OF THE HORSE.

Relation of the horse to agriculture—Time an estate—The agricultural labourer; his needs, powers, and capacities —The best men leave the soil; why—How to retain them—Our peasantry are literally serfs—Reasons for their laziness—Forced to the workhouse.

Means of rescue—The agitator a shameless humbug— Pictures of the possible—The labourer under the new system—What has been done can be done again—Effect on the poor-laws—Decrease of paupers—Lessening of rates—Workhouse expenditure—This digression useful— The horse the basis of reform . . . . 326—335

CHAPTER XXV.

DRAUGHT-HORSES AND TROTTERS.

PAGE

Farm-horses—Good thoroughbred sires for hunters—Selection—Pedigree not so essential as shape, etc.—Qualities of ancestors to be studied—Good make very necessary—Mating cart-horses.

Various draught-horses—The dray-horse—The Cleveland Bay—The Suffolk Punch—The Devonshire Packhorse—New strains—Breeding and use of all these—Intermixture of old local breeds—Gradual improvement.

The American trotting-horse—Mr. Hiram Woodruff's book about it—*Messenger*, the parent sire of the breed—His son *Top-gallant*; an extraordinary horse—Anecdote of his speed—A race in the dark—Another 'yarn'—Extraordinary fecundity of *Messenger* and his offspring—*Flora Temple* and *Dexter*; their feats—Mode of rearing trotters—My criticism on Mr. Woodruff's views—Physic and diet . . . . 336—349

CHAPTER XXVI.

THE GALLOWAY, PONY, AND ASS.

Small kinds of horses—Derivation from a common stock—The galloway—The pony—Feats of endurance; my father's boyish experience; *Sir Teddy*—Origin of small breeds—Cold climates and poor living; this questioned—Other causes may have operated—Arguments—Natural selection—Giants and dwarfs obtainable from a common source.

The ass—Various kinds—Syrian and Spanish asses—Crosses between equine and asinine species; kinship yet distinction—Swiftness of asses—The onager or kiang—Improvement of the English ass—The 'coster's moke'—Endurance—Country donkeys—Reasons for poorness of the breed—The English ass the most inferior of his kind—

Selection and treatment called for—Comparisons—Great improvement possible—A suggestion to landlords and others—What has been done for horses might be done for asses . . . . 350—363

CHAPTER XXVII.

THE STUD FARM.—THE LAND.

Choice of locality and site—Aspect—Rich pasture not essential—Varieties of soil ; examples—Ineligible sites—Cold and damp to be avoided—The land on which the best horses have been bred—Level pastures preferable.

Management of the grass—Feeding off with cattle and sheep—Proper rotation—Size of the enclosures—Whitethorn and blackthorn hedges—Superiority of live fences.

Shelter-groves around the paddocks—Selection of trees—How and when to plant—Size—Undergrowth—Species to be avoided—Hedges and railings—The wrong sort of post-and-rail, and the right—Illustrations of both—Shape of the paddocks—Trees within them . . . 364—376

CHAPTER XXVIII.

THE STUD FARM.—THE BUILDINGS.

The master's house ; central position—The men's cottages—The stables—Partitioning of the boxes—Cecil on the south aspect—My view of his theory—The roof—Collection of rain-water—Tanks and troughs—Shooting and guttering—Materials of the walls—Arrangements of interior—The manger and racks—Provision against accidents—Drains—Ventilation—Flooring—Doors.

The central yard—The various boxes—The roofing and the loft—Storage of hay, straw, and corn—Sleeping-room for stable-boys—This a necessary precaution—Boxes especially designed for stallions—Various materials—Cost of carriage—The furze hovel ; how made—General arrangement of the stallions' boxes—Enclosed yards for exercise—Concluding remarks . . . . . . 377—391

CHAPTER XXIX.

SOME ERRORS AND FALLACIES.

How able writers fall into error—'Stonehenge' on training—His mistake exposed—Buffon on interbreeding of goats and sheep—His statements fallacious—Practical experience the only safe guide—Testimony of certain flockmasters—My own observations—The question answered in the negative—Why I have gone into the subject—The thoroughly practical man always the best authority—Obsolete works on breeding and training are misleading guides—Knowledge is progressive—A mistake of Goldsmith's confuted.

'Stonehenge's' opinion on the distinctions of brain in differently bred animals—This shown to be of no practical service—A case in point—Does blood indicate quality?—Is there variation in the size and number of the bloodvessels?—These views discussed and shown to be erroneous—Does early development induce early decay of horses?—Argument and examples—I show this to be a fallacy—Proofs that our present system is a good one . . . . 392—411

CHAPTER XXX.

SOME FINAL SUGGESTIONS.

Too many stallions—Statistics of 1882—A recommendation—Startling figures—Mares also too numerous—Existing bad qualities—How to better them—Evils propagated—The French system—Bad temper—How to remedy defects—Export sales.

Statistics of 1884—Fees—Small proportion of good stallions—A warning to breeders—Mares that should be excluded from 'The Stud-Book'—Excellence not properly kept up—Injudicious foreign sales—The stallions serving in 1884—Inferiority of the majority.

Fifteen thoroughbred stallions to be selected annually
—Twelve years at the stud—Proportion of mares to
stallions—Modes of choice—How to get rid of weeds—
Evils of cheap breeding—Two cases in point—Reform desirable and necessary—The probable result . . 412—429

APPENDIX I.

THE ALVEDISTON STUD IN 1873.

1. Stallions 430
2. Brood mares and foals 430
3. Yearlings 436

APPENDIX II.

LIST OF STALLIONS.

Thoroughbred sires of 1883, and stallions serving in 1884,
in England . . . . 439—442

THE HORSE:

HOW TO BREED AND REAR HIM.

CHAPTER I.

INTRODUCTION.

My aim in writing—Need of special qualification—The question of improving the breed of our horses a national question of the hour—My experience as a breeder—List of Alvediston breeding stud—Absence of authoritative modern teaching—Excellence of theory, but need of practical knowledge—My reliance for correctness on experience alone—Existing works not entirely satisfactory—Cecil's 'Stud Farm'—'Stonehenge.'

Needful care of minor details—Value of perseverance—The method and order of the work set forth in detail.

IN attempting to do anything, there must always be an object in view. With a writer that object should be to entertain, to instruct, or in the highest sense of his calling to benefit his fellows. Under what category my present effort may rank, I do not pretend to say. But whether I succeed in the project I have formed, or whether I fail in the attempt, others will decide. To do more than add a mite to the large store of accumulated knowledge is beyond my expectation; to do less will not be for the want of effort to go beyond that on my part. Dr. Johnson tells us 'no writer pleases all, and every writer may please some.'

The world is composed of atoms. As the Arabian proverb has it, 'Drops added to drops constitute the ocean.' My contribution is but a drop added to the store of general knowledge, and as such it must be viewed. It is written without pretension, and if nothing is achieved, I must be content with the knowledge that I have escaped one evil—the blame that the ingenious author of 'The Court of Augustus' incurred, who, his critic says, knowing well the opinion of Horace concerning those that dazzle the understanding with magnificent promises, should know likewise that common-sense approves the direction of the same great authority, 'that no man should promise what he cannot perform.' This fault I have no intention of committing.

I may, therefore, present here with propriety the reason which has induced me to venture to undertake the treatment of a subject of such vast importance as the breeding not only of race-horses, but of those for the chase and other purposes; a science in which, above all other countries in the world, we so greatly excel. In the first place, for the reasons set forth in the preface, the necessity of such a work is obvious. Indeed, a powerful incentive to my making the attempt is the regrettable silence of gentlemen many of whom are pre-eminently fitted for a work of the kind. 'It is always,' says a well-known author, 'laudable to attempt much, even when the enterprise is above the strength that undertakes it.' Mine, it is true, is not a polished and elegant treatise by a distinguished essayist, nor is this what is wanted so much as the plain practical work of an unpretending expert. And such a work, to be effective for its purpose, should endeavour to embrace not only the more important matters which obviously interest the student, but also the minor incidents which in a hundred ways may teach us something of ills to be prevented or cured.

The question of improving the breed of our horses, or perhaps I should say of making the most of our already existing advantages in blood and breeding, is an important one, apart from its interest to the breeder of thoroughbred stock. Indeed, I apprehend it is a vital one to the agriculturist generally. It has for some time been clear that only in unusually favourable circumstances can farming be made to pay by growing grain. Stock-raising is gradually taking the place of arable farming. The extent of pasture is increasing yearly; we have here, therefore, the area for increasing the number of our horses, whilst that the demand for sound animals is a never-failing one is an incontestable fact.

The conclusion reached by the Royal Commission on the subject was satisfactory in one sense only; viz., that the breeding of horses is not falling off, but rather that the demand is greater than the supply. The unsatisfactory feature is that to meet this demand the right sort of horse is not bred, or is bred under such circumstances of risk as to seriously diminish the profit of the occupation. The demand, however, exists; and there should be no difficulty in supplying it, and making that supply remunerative, if the knowledge were attainable as to the best methods to be pursued by the breeders. I have convinced myself by careful inquiry that there is much to be gleaned on this subject, which I hold is of importance to the farmer before all other men, as he it is who will most profit by the information. And to bring this information to his door in a methodical manner is one of the principal aims, if not the great aim, of this book.

I have at least some pretension to essay the task I have set myself; for although I had other early employments, I have devoted a considerable portion of my time to the study of breeding horses, at a period when breeding was not thought so much of as it is to-day. And more than this, at one time, in 1873, I was perhaps the largest breeder of thoroughbred stock in England, although, a few years before, my numbers had been surpassed by those got together by the late Mr. Blenkiron in his gigantic stud at Middle Park, Eltham. As it may interest the reader, I give in an appendix a list of the brood-mares, yearlings, and stallions at Alvediston in the year named, from a printed record published at the time. It will be seen it included thirty-three yearlings, fifty-three mares, many of them with foal at foot, and three stallions. As to their quality it does not become me to speak; but I may quote the opinion of another, himself a breeder and an acknowledged judge of racing—the late Lord Ribblesdale. On the dispersion of my stud at Alexandra Park, that nobleman wrote and complimented me on having, as he said, 'the best collection of young well-bred mares he ever saw in the possession of any one man at the same time.'

Of the value of such experience as mine has been there can be, I submit, no doubt, always provided that the lesson and its application have been absorbed intelligently. I do not for a moment oppose the idea that much and valuable information may be gathered from study of early writers. Yet I think even this will not be altogether satisfactory, lacking as it does the essential part which is practical and modern, and can only be attained by constant watchfulness in the management of a large stud. There are many useful hints on breeding given by excellent men, who may or may not have kept a brood-mare or even owned a stallion, just as there are many who can write fluently and rivet the attention of their readers in amusingly recounting ideal hunting scenes, yet may never have crossed a horse or seen a hound. Still, these hints must fail to attract attention, and for all practical purposes become useless. But this is not my case. At least I can say that all I set forth as of my own knowledge I know to be correct, and that what is given on the authority of others is substantially so. I do not claim to have escaped all errors and inadvertencies. These, despite every care, will be found : ' That all our facts will be authentic, or all our remarks just, we do not venture to promise : we can relate but what we hear ; we can point out but what we see.' But even at this risk I am induced to think that the relation of what has come under my own notice may prove of service to those interested in breeding and rearing horses, from the fact that no previous writer on the subject, so far as I know, has had an equal opportunity of gaining experience by the entire control of such an extensive breeding establishment wholly his own property.

With no wish to disparage the usefulness of existing works on breeding, I feel nevertheless compelled to say that none of those that I have seen satisfy me in the matter of being in the true sense thoroughly comprehensive and exhaustive. Too often the subject is mixed up with extraneous matter, such as breaking and training the colt, the treatment of disease, or the cultivation of land for green-food—matters which could better be treated in separate works, or at least only, when necessary, incidentally referred to. Thus Cecil's 'Stud Farm,' an excellent work, and for its size containing much information, also deals with diseases, shoeing, and other matters having but little bearing on the subject proper. Admiral Rous's 'Horse-Racing' gives only a small space to the management of a breeding-stud, and advances nothing new. 'Stonehenge' deals with the subject in a masterly style. The work is by far the most complete that I have seen, and though written some years ago, it has many claims to the attention of the reader. But even 'Stonehenge' is not up to the day. 'Newmarket and Arabia,' a recent work to which I refer elsewhere, is for the reasons there given not satisfactory.

It will be seen, therefore, that I do not consider that existing works meet the requirements for which the age is ripe. To do justice to the subject, its treatment must be exhaustive, and abounding in practical knowledge, recording faithfully every known matter that may be of service in carrying on a breeding establishment with success. Small things are often of more importance to the success of great undertakings than they are generally supposed to be by superficial observers, and therefore the careful study of them should not be omitted.

Such a book as I have here outlined should, I think, prove acceptable and useful, either as supplementary to, or as replacing, existing works on breeding and rearing the various sorts of horses.

Having set forth the reasons which justify the attempt, which itself, let us hope, may be justified in the result attained, I may conveniently, in conclusion, outline the subjects which I propose to treat, and the order in which they will be treated. In a strictly professional sense, it may be, I ought to plunge at the outset into the subject of breeding, without any preliminary introduction whatever. There would, however, be something unfinished in such bald treatment of the matter. It is true, no doubt, that the frame is not necessarily part of the picture. Yet it adds to its attractions, and, indeed, is practically essential to its completeness. Thus, my own delineation of the important subject in hand will, I believe, be rendered only the more complete if introduced by an account of such matters as the history of the horse, and others which, if not absolutely essential, are germane to the chief topic. I must remember, too, that I am not addressing only the experienced, to whom such details may be trite, but a more general public, to whom the subject may prove interesting because novel.

I purpose, therefore, after reciting briefly some of the facts which urgently demand attention as matters of national policy, shortly to set forth the history of the horse; to show in the same brief way the improvement derived from the encouragement of racing, and the influence on our own horses of the introduction of the Arab strain.

The ground thus cleared for the treatment of the subject proper, the breeding of the horse of all useful kinds, I propose to deal with the influences of the climate, and the principles applicable to the selection of horses of all descriptions. Descending to special classes, it is natural that I should first deal with the thoroughbred. But the reader will not run away with the notion that the interest of this portion will be confined to those engaged in breeding blood-stock only. For, of necessity, much that will be said of the thoroughbred will be applicable to other classes also; and as space will not allow repetition, this portion should be carefully read by all those who wish to acquaint themselves with my views as to the treatment and selection of sire and dam of any kind. The selection of the mare, the selection of the stallion, and the necessary corollary, the meeting of the two, will in turn occupy our attention. Here completeness rather than brevity will be studied. In the same way will then follow the consideration of the points which should guide the breeder in the choice of dam and sire likely to give the most useful produce for the chase, the farm, the road, and other purposes.

Other important matters must then occupy our attention; for a knowledge of the general principles of breeding and of the art of selection is not all that is wanted to insure success. The true system of management must also be acquired. I purpose, therefore, to describe in detail the site and the arrangements of the ground, the method of construction and the materials used in the necessary buildings of a complete breeding establishment. On this will follow in due order an exposition of the correct system of management—or, I should perhaps say, my view of it. For this purpose we shall want a description of the technical details necessary in bringing the sexes together, and the treatment at the time, previously, and subsequently, of both sire and dam, and necessarily of the offspring, the latter from the time of its birth until its separation from its mother, and thereafter until ready for the sale-ring. Here again the information, though primarily directed to the treatment of thoroughbred stock, will have a general or extended interest in its application to other horses, and will be supplemented in the following chapters by the detail of the special treatment to be followed with other classes.

I shall then be tempted to enter into a wider subject—the possibility of reforming our national system of breeding; for that reform is needed, and, what perhaps is more to the point, practicable, I am fully convinced. And if space permit, some other cognate subjects may then appropriately be touched upon, and round off my task.

CHAPTER II.

BREEDING AS A NATIONAL QUESTION.

Absence of works on breeding horses of useful kinds—Number and value of horses other than thoroughbreds—One class taken as an example—The farmer's interest in breeding the hunter.

Hunting in old times—Hunting veterans: Mr. Radclyffe, Lord Radnor, Mr. Trelawny, and others—Stag, fox, and hare hunting—The Buckhounds—Lord Pembroke's harriers—The Prince of Wales in the field—Want of good horses.

Wide discussion of possible improvements in breeding—Opinions of Mr. Craven and Mr. Gilbey—Government views—Testimony to prevalence of injudicious selection, and to certain benefits from greater care—Our present deficiencies—The show at Newcastle: unsound stallions—Dearth of good horses in the West—Our natural advantages—No difficulty to increase numbers—The need is 'quality'—To be obtained by selection, and how—Practical, not scientific, information wanted.

I HAVE referred to the importance at the present time of the breeding of horses as an addition to our agricultural employments. Existing popular works on breeding deal chiefly with a very interesting subject—the breeding and rearing of thoroughbred stock. And though this is a matter of great importance, and one that I shall myself deal with at length, we must remember that there are other outdoor amusements besides racing, hunting being one of them; and yet of greater importance even than these are the industrial occupations of thousands in which the employment of the horse is an essential part. Probably in point of number the English thoroughbred cannot vie with any one of the following horses that are kept for specific but different purposes. To wit: hunting, carriage and cavalry horses, the shire horse, or even ponies. Therefore to the general public, and to those whose occupation is the breeding of horses, these severally and collectively have a claim on our attention the importance of which cannot be denied.

The wide spread, for example, of one popular outdoor amusement—hunting—and the occupation it gives to the many (not the least among whom is the farmer, who breeds a young horse or two suitable for the chase), cannot be questioned. That this exhilarating sport was known and indulged in at a very early period seems certain, for the first of the Persian Kings hunted. 'The Romans in Britain,' says Strutt in 'Sports and Pastimes,' 'constantly pursued the amusements best suited to the profession of a soldier, including hunting, running, leaping, swimming, and other exertions requiring strength and agility of body.' Canute the Dane, when he ascended the English throne, established laws for the punishment of game-killing of unprecedented severity; 'though he permitted the great thanes, bishops, and abbots' to hunt in forests belonging to the Crown, 'yet all unqualified persons were subjected to very heavy fines, not only for hunting, but even for disturbing the game.'

For killing a stag, a husbandman was reduced to slavery, and if a slave committed a like offence, death was his irrevocable doom. Yet this Prince, says the same author, prohibited the exercise of hunting or hawking on the Sabbath-day. The Normans had no less love of hunting than their predecessors; and it was the elder William and his successors who restricted the privileges of indulging in the sport, 'and imposed great penalties on those who presumed to destroy the game in the royal forests without a proper license.' Hunting is spoken of and recommended by James I. 'Certainly,' he says, 'bodily exercise and games are very commendable, as well for banishing of idleness, the mother of all vice, as for making the body able and durable for travell, which is very necessarie for a king.' Again, 'I cannot omit heere the hunting, namely, with running houndes, which is the most honourable and noblest sort thereof.' We read that Elizabeth (the female Nimrod) partook of the pleasures of the chase at the age of seventy-seven; and Walter, Bishop of Rochester, who lived in the thirteenth century, made hunting his sole employment, to the neglect of his ecclesiastical duties, at the age of eighty. Thomas à Becket himself might have enjoyed the sport of hunting for many years, but for his assassination in his own Cathedral.

The sport has strong claims on our attachment; for beyond the pleasure derived from the excitement in following the chase, it does more in recruiting the wasted energies and emaciated form than all the science of doctors, the never-failing nostrums of quacks, or the restorative potions of the apothecary. Of this we have sufficient proof by reference to the longevity of those who have engaged in it with more than common ardour, now no longer with us, as well as of those who are still following their favourite amusement with a zeal that partakes of passion.

That hunting the various descriptions of game must tend to longevity, no one can doubt. Instances we have in Wiltshire and the bordering county of Dorset—Mr. Radclyffe, of Hyde, who, when a young man, was *aide-de-camp* to the Duke of Wellington; Viscount Portman, of Bryanston, now in his eighty-eighth year; and amongst other noted veterans I can place Mr. T. A. Smith, of Tedworth, the late Mr. Farquharson, and the present Earl of Radnor, whose cheery heartiness to all who came near him on the hunting-field is proverbial. These gentlemen have all hunted the county, or bordering ones, for a number of years in good style and showing excellent sport. Indeed, their respective huntsmen show no exception to the general rule of the healthy and life-giving nature of the chase ; for, like their masters, Messrs. Dale, Carter, and Tredwell either lived to be, or are now, at a good old age.

We have had two or three well-known and extraordinary characters amongst veteran followers of the hounds, such as Mr. C. Trelawny, of Coldrenick, and Mr. John Russell, the sporting parson, who, the last time I had the pleasure of seeing him, was on the course of Epsom, being then in his eighty-first year, enjoying the sport in good health. And who that has ever seen him will forget the erect form of Mr. Davis in his hunting costume, riding his favourite grey up the Ascot course, heading the royal procession before the races commenced, when a very old man ? In fact, most of the above were, or are, octogenarians, if the average of their ages were not still greater.

For this and many other reasons hunting may well be spoken of in favourable terms by all who can ride a horse, and have the courage to follow the hounds, of whatever description. Some prefer hunting the deer bred and kept in large parks for the purpose, such as at Windsor.

Stag-hunting, I may perhaps be allowed to observe here, is, in its highest form, certainly the most superior kind of sport. When in North Devon or the New Forest the coverts are drawn, and a wild stag started, the pleasure of following such noble game is not to be surpassed by that afforded by any other sport that can be had within the four corners of the United Kingdom, or, for all I know, out of it. Hunting the park-bred deer, uncarted in sight of the field, with the hounds laid on the scent after the usual law, has by comparison a certain amount of tameness, akin to that experienced by the frequenter of the shires when put to hunt 'a bagman' with fox-hounds. Nevertheless, it has its own merits : it affords sport to those who otherwise could not find it in distant scenes. A good gallop is often the result, and at least it precludes the catastrophe of a blank day. And beyond this, in the special case of her Majesty's Buckhounds, most excellent sport is to be had ; for under the able guidance of the Earl of Coventry good runs are the rule, rather than the exception.

Hare-hunting finds its votaries in localities unsuited to fox-hunting, or where the fox is not to be found, and has a special advantage in affording the pleasures of the chase to those who, from age or other reasons, are not willing or able to ride hard. At times harriers afford excellent sport, as was often witnessed with the pack lately hunted by the Earl of Pembroke, who is likely, I believe, to become an M.F.H. before long ; for which occupation, I need scarcely say, his genial liking for the sport, and his qualifications as a bold and judicious rider, eminently fit him.

Hunting, indeed, like racing, has been for generations the sport of kings and their august consorts. In our own time, the Heir Apparent to the throne shows no forgetfulness of these splendid traditions, whilst he exhibits every capacity for filling them. The Prince of Wales is fondly attached to fox-hunting, and rides nothing but wellbred horses. He is not only a good horseman, having an elegant seat, but is also an excellent judge of the sport; and few of the nobility, or, indeed, of any other class, can ride across country with firmer nerve or better judgment, as I have often had occasion to observe when seeing his Royal Highness out with the hounds. Fox-hunting, which is only equalled in excitement by the chase of the wild stag, is the most popular form of the sport, and is indulged in by all classes of society that can afford the time and the money for its pursuit. It is, in short, one amongst the many causes which have created the demand for a good and capable horse, the supply of which should, I am convinced, remuneratively employ our yeomen and farmers.

There has not certainly in my time, or perhaps in the memory of the oldest man living, been so much written or said on breeding horses for the cavalry, hunters, and the carriage, as within these last few years, raising a discussion much the same as that which took place on the introduction of foreign horses for the improvement of thoroughbred stock, about the year 1750, or rather over a century ago. That this latter ended successfully, no one will pretend to doubt; nor, I think, will any have misgivings as to the advantage of the present discussion on breeding all descriptions of horses for our present and future use, whether in time of peace or war. Many shrewd men of business have given us the benefit of their practical experience. Learned treatises have been written by able and conscientious men, each holding different views. One gentleman assures us that it is incumbent on us to breed more horses; and others, seemingly with more likelihood, assert that it is better horses that are required, and not an increase in number.

Only a few months ago the matter was discussed in an able and business-like way by Mr. W. G. Craven, in a letter to the *Daily Telegraph*. He thinks that the breeding of horses for the cavalry, artillery, and transport service should be taken up by the Government. This is a matter, no doubt, to which he has given much attention, and from his experience in thoroughbreds and other descriptions of horses such an opinion is valuable. But these views are diametrically opposed by the Government. In reply to Colonel Hughes-Hallett's question, whether the Government would take into serious consideration the advisability of checking for a time the export of horses from this country—at all events, until the requirements of the artillery and cavalry were fully satisfied—Mr. W. H. Smith said in substance, 'Her Majesty's Government have no idea of doing such a thing; for,' he added, 'there is no reason to believe either that the present exportation has assumed serious dimensions, or that the army is short of its authorized establishment of horses.'

Whether it is a question that the Government alone ought to grapple with on an extensive scale, or one that would be more favourably carried out by private enterprise, is a matter on which there are different opinions. Mr. Walter Gilbey, who has done much towards the improvement and raising the value of the shire and other horses, says it is not State aid or Government breeding establishments on a large scale that are wanted, as he thinks private enterprise is sufficient for breeding horses of all descriptions for our own use. This, indeed, I may say, is, in a sense, confirmed by Mr. W. G. Craven, who expresses a further opinion that there is plenty of room for the formation of other studs than for thoroughbred stock. Here, I have no manner of doubt, and for the reason he gives, that he is right. He also wisely deprecates the use of stallions that are not in every way suited for the purpose. No one can question the soundness of this policy, for the evils of its contravention are unfortunately to be seen every day of one's life, unless indeed we are wilfully blind. This being so with thoroughbred stock, one can hardly see any reason why it should not be so with other descriptions of horses. It is therefore, in my opinion, not that we require more mares to breed from or horses to serve them, but rather a better description of sire and dam throughout the country. If we wish to retain our national character as breeders of the best horses of every description in the world, we must have more regard and pay greater attention to the selection of the stock that we breed from, or we shall soon be passed and beaten in the race for supremacy by the enterprising foreigner, and we shall then have to buy of them instead of, as now, having them come to us for the best horses from which to breed. There should really be as much judgment used in selecting a hunting mare and a thoroughbred stallion to mate her with, or the sire and dam of a good draught-horse, as there should be in choosing the stock from which to breed a race-horse; for, as Mr. Walter Gilbey very properly says, 'The sight of fine teams of them (cart-horses) fills the mind with a notion that there is hardly anything which cannot be done in the way of modifying size and form by careful and prolonged attention to the science of breeding.'

How far we are at present from realizing this desirable condition may be gathered from the following :

Commenting upon the Hunter and Stallion Show of the Royal Agricultural Society of England, lately held at Newcastle, which a reviewer describes as 'one of the best collections of thoroughbred stallions, to judge them by appearance, ever seen,' he observes, and I have no doubt with candour and ability, that, 'out of the thirty-six stallions exhibited, no less than twenty-four of the number were more or less disliked ; ten were condemned for having bad feet or legs, or defective hocks' (which, to my thinking, is the greatest of all evils ; for if horses are not well qualified in these respects, they may possess all other good points and yet be perfectly useless) ; 'two were cast presumably for unsoundness, and two more were hopelessly suffering from the same affliction.' And how many others would have been placed in the same category, if they had all been subjected to a trying ordeal at the hands of the veterinary examiners, no one can tell ; but, I suspect, not a few. Here we have, out of thirty-six picked stallions, supposed to be the best of the sort in England for getting hunters and carriage-horses, no less than fourteen, or nearly half (perhaps more if they had all been examined), considered by judges not fit for the purpose. This is apart from the question of their soundness in respect to 'roaring;' 'but report does add,' says the reviewer, 'that several moderate-looking thoroughbreds have also travelled, and that the half-bred stallions outnumber the thoroughbreds.' This certainly cannot be a very flattering prospect for breeders in the North, the supposed cradle of horseflesh; for, moreover, we read that 'the dearth of good mares was bewailed at the show.'

It was but lately a well-known dealer in horses told me that he had discontinued his visits to the West, where formerly he used to procure most of his hunters and carriage-horses, as there were none to be had; for now only a few are bred there, and these disposed of at an earlier age than his customers cared to have them. I think the same remarks, if I am not greatly mistaken, will, with as much truth, apply to other localities that formerly were the great centres from which hunters, carriage-horses, and other descriptions of useful animals could be procured. Still, there are many counties that breed horses more abundantly than they did formerly, and in computing the numbers that are now annually bred, this must be taken into consideration. Indeed, I hear on good authority that in Devonshire there has been lately a great revival among the farmers and others in breeding good half-bred stock.

This I can understand. There is, in truth, no difficulty in increasing our number of horses—

almost without limit, I might say. Why, therefore, should we not breed our own horses in sufficient number to meet all requirements for every purpose? I am convinced that we have plenty of parent stock in the country—horses and mares—for the purpose. And from all the natural advantages of the climate, the cheapness of the land (owing to the low price of corn), and the immense number of acres that have been turned from arable into permanent pasture, horses may be raised at a price that will defy foreign competition.

I expect that the reader will agree with me that these facts are of pre-eminent importance at the present time. Undoubtedly what is wanted is to improve the quality of our horses—not in individual instances, which I am happy to think it would scarcely be possible to do, but all along the line. With attention to this essential we could have, I am convinced, a better selection from native produce at a cheaper rate, especially if quality be taken into consideration, as it ought and most assuredly would be. In contemplation of these facts, I think that the conclusion reached by all sensible people will be that we should extend the business of breeding horses of all classes, so as to be independent of extraneous aid. Whether the matter be carried out by public companies, by Government undertakings, or preferably, as I think, by private enterprise, is immaterial. What is clear is that farmers and others would greatly assist by breeding, and breeding with care, horses of different kinds, of which the cavalry horse is one example; and that, too, for their own benefit, as, taking the particular class alluded to, horses never were dearer, nor in greater demand.

All that has been said indicates the absolute necessity, in order to achieve real success, of carefully selecting both your sire and dam, and seeing that they themselves are descended from a good stock. To help the breeder to this desirable end is the primary object that I have in view. To be serviceable, the description of the kind of horses that should be selected need not be scientific, but given in a manner that may easily be understood and followed with the least trouble and expense, even by young beginners. Every man may be his own gardener, or his own lawyer (save us from him!), and anyone may be the breeder of his own horses; nor need he despair of success, if he will only, in the first instance, pay attention (though it must be careful attention) to the rules of common-sense in selecting his stock. And these pages, I trust, will help him to that desirable end.

CHAPTER III.

HISTORY OF THE HORSE.

Numerous authorities on the subject—Prehistoric remains—Zoological classification—General description—Always much valued—Original habitat and gradual extension—Early use in Egypt—Biblical testimony—Claim of Arabia as its origin examined—My own negative opinion confirmed—Use in Persia and Palestine.

The horse in Britain—Records of large animals in a wild state—Used in the invasion by Cæsar—Cæsar's testimony to its previous subjugation in Britain—Already an animal of value—Of necessity of mixed breed—First trace of distinct crosses.

THE history of the horse, from the very earliest period in which one finds any reliable account of his having been domesticated and made subservient to the use of man, or of his being brought to contribute (as he does) in various ways to man's many pleasures, has been written by numerous writers of great eminence. These histories form a chain of evidence that can leave no doubt on the mind of attentive readers that in the main they are substantially correct, being corroborated in so many ways by different writers at greatly differing distances of time. To quote from all, or name more than a few of such chroniclers, would be unnecessary, and to leave these records untouched would be to render my brief account deficient in matters of ancient history.

The antiquity of the horse is very great, for it is said he can be traced back to the earliest tertiary age, but we have no cognizance of any mammals of the group to which he belongs before the days of the eocene period. We are told that at that time his diminutive form, or that of an animal resembling a horse, was not bigger than that of a fox. In the miocene period it became as large as a sheep, and in the pliocene time was the size of a modern donkey, but it was not till the pleistocene period that equidæ appeared which approached the size of the existing horse. 'When I found in La Plata,' says Mr. Darwin, 'the tooth of a horse embedded with the remains of a mastodon, a megatherium, a toxodon, and other extinct monsters, I was filled with astonishment, but my astonishment was groundless. Professor Owen soon perceived that the tooth, though so like that of the existing horse, belongs to an extinct species.' If it be added that fossil remains of true horses, differing but very slightly from the smaller and inferior breeds of those now existing, are found abundantly in deposits of the most recent geological age in almost every part of America, we have proof of the existence of some creature akin to the horse of our own time, in some unknown shape, in the remotest ages. This is, however, but a matter by the way; what is of more interest at present is the history and description of the animal as he exists and is known to us.

Zoologists class the horse with the mammalia, the first grand division of vertebrate animals, which is placed at the head of the animal kingdom. Cuvier considers that one genus only is comprehended in this group or family, viz. *Equus*, which he places in the 'solipedous section of the *Pachydermata*, quadrupeds which have only a single toe on each limb apparent, incased in a hoof, although there are on each side of the metacarpus and metatarsus stylets which represent two lateral toes.' Gray, on the other hand, considers the *Equidæ* to consist of two genera, '*Equus* and *Asinus*.' Many modern zoologists have adopted his views. One authority (Colonel H. Smith) separates from *Equus* and *Asinus* those species which are striped like the zebra, applying to them the name of *Hippotigris*. Other eminent zoologists consider that 'the horse and its near allies, the several species of ass and zebra which constitute the genus *Equus*, comprise at the present time six types, sufficiently distinct to be reckoned as species by all zoologists, and easily distinguished by their external character.'

I do not profess in the above to have done more than give at second-hand facts already known to the student. A brief recital of them, however, could not well have been omitted, in order to define the accepted opinion of the genus of the animal. As for the existing varieties of the domesticated horse, I cannot, I think, do better by way of description than once more transcribe that of a well-known authority:

'The horse (*Equus caballus*),' says this writer, 'in a state of domestication, varies in size from the massive and gigantic dray-horse to the diminutive Shetland pony. Its colour is also as variable as its size, ranging through white, cream-tint, gray, mottled, iron-gray, dun, bay, chestnut, black, etc. It, moreover, presents us with different strains, how produced originally is not easy to say: certainly the high-blood racer and the slight meagre Arab present strong contrast in their contour and capabilities to the heavy Flanders horse and the huge dray-horse, rising from 18 to 20 hands high at the withers. Between these two extremes there are numerous intermediate breeds: some adapted for the chase; some for carriage and light-wheeled vehicles; some for the traveller's saddle; and some for farm-labour. Ponies, again, a small variety of the horse, show differences of a like nature: some are fine-framed, full of mettle and courage, and show high blood; others are clumsy and ill-formed, though strong and hardy; and the intermediate gradations are numerous.'

I must be content with this brief allusion to the natural history of the horse as prefatory to more important matters. I cannot, however, refrain from confirming in my own way what has been said, by observing that this noble creature has been spoken of from the earliest date by the ancients with something like affectionate veneration, and to the present day has lost none of his claims to our protection and particular notice. Indeed, he is looked upon with admiration, and held in increased regard, whether we take the race-horse, doubtless in symmetry and swiftness, as well as for endurance, the most exalted and beautiful of all his race; the patient, plodding cart-horse, or others that form continuous links, down even to the most diminutive of the tribe—the hardy Shetland pony. All and each are severally fitted in the most perfect way for the different parts they have to perform.

If I have relied on recognised authorities for the natural history of the horse, I shall, in tracing its origin and habitat, write independently. For one thing, I shall go for important facts to the Sacred Volume itself, premising that I do so with the most profound reverence, entertaining the utmost veneration for its sublime truths. This may have its disadvantages. Scepticism is accepted by some, but I trust by the unthinking, as an easy proof of superior intelligence. But this is not so, I am happy to believe, with all; and to many it will, I apprehend, be a satisfaction to find that my researches unmistakably prove that the spread of the horse east and west, north and south, from a certain centre, and its non-existence in the New World, only go to confirm the Biblical account. We have evidence of the early existence of the horse in Persia and Armenia, in which countries it would first be propagated after its liberation from the Ark. From this centre it would appear to have found its way to Egypt. Pharaoh appears as the first possessor of this noble creature in large numbers. This is easily seen by reference to early writers both of sacred and profane history. He probably used it for pleasure as well as for war, and for that pompous show which the people in the East indulge in now with as much vanity as in primitive times.

The glowing account of the horse as given in Biblical history at a very early date, in the Book of Job, surpasses in descriptive beauty, poetical metaphor, and sublime language, all that has before or since been written on the subject. There can be no doubt that Job lived in patriarchal days in the land of Uz, in Arabia, and that the country was not far from Egypt. But in that vivid description of the horse he is not said by the sacred historian to be a native of Arabia, nor, indeed, is there any other habitat assigned to him. He may, from anything there said to the contrary, have been from Egypt, and most likely he was; for it is there he is first mentioned in Holy Scripture as being subjugated, in the time of Joseph. Commentators of great ability reckon this period to have been prior to the days of Job, who lived, according to their computation of time, soon after Joseph and before, or in, the days of Moses. In corroboration of this we know that 'Joseph sent waggons' (interpreted by the Rev. William Owens as 'chariots drawn by horses') for his father, brethren, and their families to Hebron, in the land of Canaan, to convey them to Egypt, when on their arrival in a strange land Joseph, in his chariot, drove to Goshen to meet them. On this point I may perhaps refer to the notes of the Rev. Thos. Scott on Genesis xlv. 19. He remarks: 'No mention has hitherto been made of horses amongst the possessions of the patriarchs, or of wheeled carriages, both of which abounded in Egypt at that time; it is probable these waggons were drawn by horses.'

Here is apparent proof that the Egyptians had horses in those days, and most likely in large numbers; for soon after we read that Joseph gave the Egyptians food in exchange for their horses. Shortly after this event horses and chariots and a very great company were at Jacob's funeral, when he was taken to the land of Canaan to be buried in the cave of Ephron the Hittite. And in that stupendous miracle, the overthrow of Pharaoh, 'a new king over Egypt which knew not Joseph,' we read that his host, his six hundred chosen chariots, and all the chariots of Egypt and horsemen, were overwhelmed in the Red Sea. Later, it is said on the same unerring authority that 'King Solomon got his horses from Egypt,' for whose accommodation, we read, he had forty thousand stalls.

These facts lead directly to the consideration of the allegation popularly put forward, that Arabia must be regarded as the original habitat of the horse. I have already observed that in the description found in Job, no one locality is fixed upon as its place of origin; and that the earliest mention of its subjugation shows it to have been domesticated in Egypt. And now we find that King Solomon collected his vast concourse of horses from Egypt. I think it may fairly be asked, why should he not have done so from Arabia, if they existed there in great numbers at the time? Mr. Bell confirms this opinion in his 'History of British Quadrupeds.' Speaking of the Arabian horse, he says: 'There is no proof that it was indigenous to that arid country, for there is great reason to conclude that it was only at a comparatively late period that it was employed by that people' (the Arabians); and adds that 'there appears a great probability in the opinion that Egypt or its neighbourhood' (he does not say

Arabia) 'is its original country, and still more that this extraordinary people first rendered it subservient to man, and subsequently distributed it to other countries.' The period that elapsed from the Deluge to the death of Joseph was, according to the calculation of the Rev. Thomas Scott, seven hundred and thirteen years. We can, therefore, readily believe what we read of the great numbers existing in Egypt at the later date. For instance, a single pair would have propagated thousands and tens of thousands, a number, indeed, beyond computation, if we remember that they were running wild; for it is not until Joseph's time that we hear of their being domesticated, though probably before that period the Egyptians had horses in subjugation. How, then, can any individual pretend to prescribe limits within which such a vast multitude would range when in search of food, or impelled by climatic circumstances to seek different countries, or by the still more powerful incentive to their roving dispositions, the natural desire of propagating their species? There may at the time have been horses in Arabia as well as in Egypt, and probably also in many other bordering or distant countries.

I have said so much on this point because we find some authors who will undertake to say, and essay to prove, that the horses of Egypt were not mixed with the Arabians, or *vice versâ*, or, indeed, with others of their own species, from whatever part, however distant they may have come. Ancient history affords no proof whatever that the horse was indigenous to Arabia prior to the fifth century. That he there existed I do not doubt, but was probably, before captivity, of a mixed herd from many countries. Horses, in fact, were caught wild and domesticated in Arabia as in other countries, and could in no sense have been a pure breed. This is corroborated by the fact that 'Mohammed, during his early career, was badly off for horses, and could not possibly capture them,' from which I infer that he had not the necessary means to do so; or, what is still more likely, that they were scarce and difficult to find. 'It would appear,' says the latest authority that I have consulted, 'that the horse was first domesticated or reclaimed in the East; that it was brought with the hordes migrating westwards from Asia, and was thus introduced into Arabia and Egypt, and that we must look to the deserts north of Hindostan and Persia as its cradle, or, at least, for the locality in which it first became subject to man.' This view largely confirms my own, and that is, that we must not look to Arabia as the original habitat, but rather to Central Asia, whence the horse passed into Egypt, and there acquired its first importance as the friend and servant of man.

That the horse existed in Persia at an early

date I cannot doubt. That King Cyrus kept hunters for his amusement is certain. But this was at a later period, as also was the use of the horse for warlike purposes by the Romans in Palestine. It is recorded in Josephus's 'History of the Jewish Wars' that the Roman Emperor Vespasian sent one thousand horsemen and two thousand footmen to attack Japha, a city that lay near to Jotapata, which was taken and destroyed by Trajan, the commander of the Tenth Legion. I mention this as illustrating the use of the horse for war purposes in the East; but, as a matter of fact, this incident was later in date than his known use for the purpose, and existence in subjugation, in Britain, to which I must now refer.

It is by no means clear whether or not the horse was indigenous to this country. Certainly, he is not alluded to by any naturalist as existing in a wild state. Cattle, bears, pigs, and wolves are the only large animals mentioned by the early writers to whom I have access. Earlier species, no doubt, existed, and have been duly classified from fossil remains. Fitz-Stephen, who wrote in the reign of Edward II., tells us that 'the forest by which London was then surrounded was frequented by boars as well as by various other wild animals.' Bewick says the bear (*Ursus arctos*) was once an inhabitant of this island, and was included in the ancient laws and regulations respecting hunting. Dr. Fleming observes of the indigenous bear of Britain : ' These animals not only prevailed in this country at the period of the Roman invasion, for Plutarch relates that they were transported to Rome, but maintained their existence, in spite of the efforts of the huntsman, to the middle of the eleventh century at least. Cattle in a wild state certainly tenanted the British plains anterior to the earliest known records. Vast herds also remained unreclaimed to a comparatively late period, and several varieties, if not a distinct species, of the *Bos taurus* formerly existed amongst us in a wild state.'

With regard to the existence of the horse itself in Britain, we find conclusive evidence in that earliest authentic history of our country, written by Julius Cæsar in the fifth book of his ' Gallic Wars.' It is certain he brought horses with him in his second descent upon our shores, if not on his first invasion ; for he observes of one of the early encounters with the natives that ' his cavalry drove them (the Britons) into the wood in rear of their position.' Nevertheless, though the Romans may have added variety or increase to the number of horses already existing in the island, it is equally certain that they had been introduced into it at an earlier period, though there is nothing to tell us by whom or at what time this was done. This we have on the evidence of Cæsar himself, who, in the same account, goes on to relate that soon after he withdrew his troops, and on his subsequent return to his former post, found the native princes had augmented and combined their several forces under the commander-in-chief, Cassivelaunus, to defend themselves from the attack of their common enemy, and for the purpose of expelling him from their native shores; that after some severe and unsuccessful fighting, the British prince was obliged to dismiss the greater part of his forces, retaining about four thousand charioteers. From this it is positively certain that the horse must not only have been here, but plentiful in the island, before and on the arrival of Cæsar, B.C. 55. The Britons are described by him as a warlike people, fighting on foot, on horseback, and in chariots. From accounts given of, and blades that have been dug up on, ancient battle-fields, the chariots seem to have been armed with scythes attached to the axle-trees; thus confirming the opinion that horses were in Britain, and the art of horsemanship—though probably in a rude way—had been learnt and practised before the invasion of Cæsar. Indeed, it would seem that the animal, as it then existed with us, was already a good specimen of its kind, for we read in Collier's 'British Empire' that 'in the time of the Romans British cattle, horses, and dogs were much prized.' The horse may have been an aboriginal native of the country; probably he was, in the same way as we find that cattle were known in a wild state. The evidence, however, so far as it goes, is negative ; for, as I have observed, the horse is not mentioned by naturalists amongst the wild animals. What we do know is, that at the time of Cæsar's invasion he not only existed, but was used in warfare by the Britons, and was, even at that early age, thought to be of considerable value. But, whatever his qualities, the horse was, as in other parts, soon mixed in blood with the horses of different nations. From this point we can begin to trace distinct crosses with other kinds, a subject which will be best examined when we come to consider the question of breeding in times gone by.

CHAPTER IV.

PROGRESS OF RACING, AND ITS RESULTS.

Decline of cruelty in sport—Corresponding increase in racing—Earliest records—Homer—Racing colts in Greece.

Racing in England—In Athelstan's time—Henry II.—Henry VIII. and Elizabeth—Encouragement given to breeding by the Stuarts—James I. imports horses—The *Markham Arabian*—Cromwell's *White Turk*—Charles II. a genuine benefactor—The royal mares—Commencement of the Stud-book—Racing at Newmarket—Queen Anne—Importation of noted Eastern sires—George IV.—Anecdote of the 'Sailor King'—H.R.H. the Prince of Wales as a breeder and owner—The Hampton Court Stud—Foreign triumphs in breeding only exceptional.

THE progress of civilization in this country has been marked in no happier way than by the gradual relinquishment of cruelty in our national sports. The days of prize-fights, bull and badger baiting, and of dog-fighting, are not only at an end, but are hardly in the remembrance of the present generation. Cock-fighting, scarcely known in our days, had at one time a great hold on the aristocracy of this country. As a sport, it has the sanction of high antiquity for its practice. It was known at a very early date in Asia and China, and was indulged in by both Greeks and Romans. Tradition says that it was introduced into this country by King Charles II., and was eagerly resorted to by all ranks of society. But it, too, has had its day, and no longer exists, except covertly, in isolated cases, in defiance of the law. The days are changed indeed since good Queen Bess—as she is styled by her admiring chroniclers—hunted the stag, which was shot at with arrows during this curious chase; or when, on her sister's visit to her at Hatfield, she was entertained with a grand baiting of the bear. This latter misnamed 'sport'—the torturing of a helpless purblind bear—was, we are told, exceedingly enjoyed by the fine ladies and Court beauties as a Sunday afternoon recreation.

Such scenes are ended, whilst another and nobler amusement, properly termed 'the sport of kings'—racing—has not only held its own from time immemorial, but has increased in favour, and for centuries has been looked on and recognised as the national sport of old England, and is now indulged in by the inhabitants of the four quarters of the habitable globe. Admirers of the sport may be found in nearly every town, village, and hamlet, through the length and breadth of the land. Moreover, renowned foreigners from nearly all quarters of the earth come to witness its exhibition for pleasure or speculation, or in other ways take an interested part in it.

Horses, no doubt, must have been bred before they could be raced. But though breeding came before racing, the latter will be glanced at first, simply to show briefly how far its practice may have influenced the breed of horses for good. We are told that 'probably the earliest instance of horse-racing in literature occurs in Homer, when the various incidents of the chariot-races at the funeral games held in honour of Patroclus, which were of a semi-religious character, are related.' That it is of very ancient date will be readily admitted. Racing was practised by the Greeks, as is mentioned by Sophocles, who gives an account of such sport at Elis, when both foot-races and equestrian trials of swiftness were made and chariot-races held. This was about six hundred years before the Christian era. Grote, in his 'History of Greece,' speaks of races 'between colts of the same nature as full-grown horses.' These I take to be yearlings or two-year-olds. At any rate, they must have been horses very much younger than those that generally raced. This passage was probably in the late Mr. Frail's mind when he first introduced the yearling race at Shrewsbury, and served him as a justification for such an innovation on the English race-course, which, however, was soon very properly prohibited by the Jockey Club.

As for racing in England, the earliest record of it that I can light upon is that given by

Strutt in his 'Sports and Pastimes of the People of England.' 'Racing, or something like it, was set going in Athelstan's reign.' We know further that this king (Athelstan) received as a present from Germany several 'running horses,' evidently race-horses. It no doubt in its earlier stages owed much of its vitality to the countenance given to it by successive monarchs. Probably most of them took an interest in it in some shape. But so far as any record is concerned, we must pass from the time of the Saxon Kings to that of the first Plantagenet. Fitz-Stephen, in his description of London at that time (A.D. 1154), says: 'Smithfield is a field where every Friday there is a celebrated rendezvous of fine horses brought hither to be sold.' He then speaks of racing, and adds that here it was first known in England. The 'strong and fleet' apparently were only allowed to contend, as 'the common horses were ordered out of the way' for the purpose of clearing the course. I presume they raced in those days for honour, and the jockey rode for applause, as no mention is made of stipulated fees or gratuities to the riders. But soon after the twelfth century racing was more common, and then they ran for stakes—'forty pounds of redy goldie;' the distance is stated as 'three miles,' and the scene in the Metropolis transferred from Smithfield to Hyde Park. This account is fully confirmed by later writers, and is sufficient to show the origin and the place of our first regular races, though it appears from the account of Mr. Cheyne, who preceded Messrs. Weatherby as a compiler of turf statistics before and between 1721-27, that there were no regular accounts kept of how the horses came in. Coming next to the time of Henry VIII., an account of his horses is given in a paper relating to the household expenditure as follows : 'Coursers, young horses, hunting geldings, hobies, Barbary horses, stallions, geldings, bottles, mail, pack, robe, and stalking horses.' That his Majesty ran horses seems certain, for a little later on we find among other items it is thus recorded 'by way of rewardes': A reward also 'to the boye that ranne the horse.' 'His chief outdoor amusements were shooting at the rounds, hunting, and horse-racing.'

Writing of this monarch, the Venetian Ambassador says: 'About the year 1521, when Henry was twenty-eight years old, he was an admirable horseman, uncommonly fond of the chase, and never engaged in it without tiring eight or ten horses.' Such a feat in horsemanship in our days would hardly be considered a test of a good rider, but rather of a hard and heavy one.

As regards racing in the time of Elizabeth, I find it recorded in Collier's 'British Empire' that 'there were horse-races for prizes ; but the modern system of gambling bets was unknown.' It is not until we come to the time of the Stuarts that we can find a proof of really steady encouragement of the national pastime, with the distinct view of improving the breed of our horses. To James I., about the year 1603, the early improvement of the English horse is undoubtedly due; for it was this monarch who, history informs us, introduced horses from the East. And they were the most famous he could find, for he bought of a merchant named Markham an Arabian horse for £500, which in those days must have been considered an enormous sum. But he was a failure both as a race-horse and at the stud, as many and most of the Arabians were then and have been since. The Duke of Newcastle, who seemed to dislike Mr. Markham's Arabian in particular and the breed generally, heightened this prejudice by his remarks in his work on horsemanship, in the second volume of which he says, 'But we have of late years run too much into the Barb and Arabian kind.' He admits they have size, but they lack substance to carry weight. How strange it is that the Duke should be the only historian of the time who says the Arabian horses have size! Most others say he is under 14 hands, or about an inch higher, whilst the Turkish horse is said by Goldsmith to be 16 hands high.

Naturally, sport of all kinds suffered during the turbulent and unhappy reign of Charles I., though races, it appears, were held at several places, and his Majesty attended. Cromwell's interdict afterwards stopped the sport for a time, although he was desirous of improving the breed of horses, owning, as he once did, the celebrated *Coffin* mare, that was found hid in a cellar, and a stallion, *Place's White Turk*. The importation of foreign horses continued during the reign of Charles I., during the time of the Commonwealth, and also during the respective reigns of Charles II. and James II.; and these arrivals, it appears, were mostly from Barbary or Turkey, and supposed to be the lineal descendants of the paragons of the arid deserts of Arabia. We cannot, however, be quite sure of this, for most foreign horses were so described by the wise men of the East, and as such believed in by their less astute brethren of the West. Amongst these horses came *Place's White Turk*, above mentioned as the property of the Protector, *Helmsley Turk*, and others of similar descent.

The reign of the Merry Monarch is not credited with much good. The lover of the horse must, however, retain some respect for the memory of Charles II., from the fact that on the Restoration he did more genuine benefit perhaps to the breed of the English race-horse than any person who preceded him. For we read that he sent his Master of the Horse abroad to purchase foreign mares of the best and purest blood, as well as stallions. The former were called, and are to this day known in the 'Stud-Book' as royal mares (though it appears little was known of their pedigree then or now), the celebrated *Eclipse* and *High Flyer* being bred in a maternal line from one or other of them, as is duly authenticated by the same authority.

Beyond doing more than any of his predecessors towards improving the breed of horses, Charles II. pursued the sport itself with keen pleasure. It is of his time that one author says : ' They had in those days nice hacks, and rode in with the race-horses to the finish at Newmarket ;' a practice allowed even in my day, and in which one may have indulged with impunity, but now prohibited, and its transgression punishable with a fine. The influence of his example did not die with him ; for Queen Anne, on ascending the throne, gave additional plates to be run for. Her Majesty both kept and ran horses ; but whether she bred any or not, I see no record. A curious circumstance we find connected with her racing was that on the very day before she died she won a plate at York with a horse named *Star*. It was run in four heats. The effect of the impulse given by Charles II.'s encouragement to breeding is shown in the fact that, in the years just preceding and subsequent to good Queen Anne's reign, viz., between

1689 and 1730, the most notable of our Eastern sires were imported, viz., *the Byerly Turk, the Darley Arabian, the Curwen Barb,* and last but best of all, *the Godolphin Arabian.* To these, and to some of the royal mares, I think we are most indebted for the improvement of our horses. That they were capable of improvement I do not doubt, for they could not have been at that time very good, although, according to a well-informed contemporary historian, they were much better than is generally believed in the present day, if they were not actually better than the Barbs and Arabians themselves.

With the accession of the House of Hanover the personal interest in racing as an amusement declined, although a generous support was accorded to it. George I., though not fond of the sport, instituted the King's Plates. George II. and George III. showed their interest in the pastime by generously subsidizing it, for the purpose of improving the breed, the increase in the number of the King's Plates being due to them. George IV., however, took a lively interest in racing. When Prince of Wales he ran many horses, and after he came to the throne he was also a breeder of horses and the founder of the Hampton Court Stud, where his brood-mares and stallions were kept till his death. William IV., 'The Sailor King,' was fond of the amusement; but, like some of his predecessors, knew nothing about horses, although desirous of patronizing the sport and improving the breed. The following characteristic anecdote is told of him. Shortly after his accession, in the June of 1830, Edwards the trainer approached his Majesty, and inquired what horses were to go to Goodwood. The King replied in nautical terms: 'Take the whole fleet; some of them, I suppose, will win.' The three horses belonging to his Majesty which were engaged in the Goodwood Cup were accordingly despatched to the scene of action, and finished as follows :

GOODWOOD CUP, AUGUST 11, 1830.

His Majesty's b. m. *Fleur-de-lis*, aged, 9 st. 9 lb. . . Geo. Nelson 1
His Majesty's b. h. *Zinganee*, 5 yrs. old, 9 st. 10 lb. . . J. Day 2
His Majesty's ch. h. *The Colonel*, 5 yrs. old, 10 st. . Pavis 3

Six other starters not placed.

A pretty good proof that the royal trainer knew his business.

In concluding this account of royal patronage of racing, it is pleasant to be able to record the fact that the Heir Apparent to the throne, H.R.H. the Prince of Wales, not only races in his own name, but has commenced a small breeding stud also at Sandringham. It is not improbable that in the end this may be the beginning of a larger undertaking, to be carried on at the royal stud at Hampton Court, which has been renewed and reformed in the present reign. And if so, it would assist, let us hope, in maintaining that proud supremacy which we have so honourably gained of breeding the best horses in the world. In saying this, I do not forget the few exceptionally good horses we have seen from France, Austria, and America—*Gladiateur*, *Kisber*, and *Iroquois*, to wit, all winners of the Derby ; and specially *Foxhall*, perhaps the best of the four, and certainly far superior to any horse of his year. These, nevertheless, when all is said, are direct descendants of our own horses, and are but exceptional, almost solitary cases, in comparison with the great number of horses bred in foreign countries.

CHAPTER V.

ORIGIN OF EXISTING BREEDS.

Sources of information—Earliest crosses—English horse valued in Saxon era—Spanish strain—Horses in 'The Stud-Book' in 1793 —The *Byerly* and *Darley Arabians*—*Curwen's Bay Barb* and its produce—*Flying Childers'* fabled speed—The *Godolphin Arabian;* no pedigree—Description—His first produce—Story of the black cat—Result of the crosses; three famous sires: *Matchem, Herod,* and *Eclipse*—Performances of the latter—*Marske*— *Highflyer*—The successors of the three in our own day.

In the brief history of the horse we traced it from the earliest record of its origin down to its domiciliation in this country. This was a point at which it was well to leave the matter of history and turn to the question of breeding. So far as I know, we have no ancient treatise on breeding. For we cannot say whether the elaborate work on the horse written about 300 B.C. by Xenophon contained any reference to the subject or not. In fact, as regards the earliest stages of the procedure, much must remain unknown, and perhaps more uncertain. When, however, we get to the period embracing the early part of the last century, we come upon more certain

4—2

ground. This period commences with the introduction into this country of the Barbs and Arabians for the purpose of crossing them with the mares we then possessed. Subsequently we come to more precise information in the epoch in which we derive our knowledge of the breed and capabilities of the horses named from the 'Stud-Book' and 'Racing Calendar.' In the last fifty years, or I may say from the year 1830, our knowledge is still more accurate. For in this period we can consider the qualities of horses and mares that have been under contemporary observation, or at least within the memory of living authorities, such information being not only more comprehensive and varied, but of greater importance to the subject of our inquiry. For we are able to trace the pedigrees and performances of different horses with accuracy; and we also have the knowledge of many peculiarities of the different animals, both sire and dam, in cases of all descriptions. Their size, whether great or small, is made familiar to many of us—a matter of no little importance; as well as their particular temperament, action, and capabilities, facts from which alone we can hope to gain real knowledge and reap substantial benefit in the study of the intricate subject of breeding.

The first trace that I can find of the crossing of our horses with any other strain of blood was after the subjugation of the Britons by the Romans.

The latter, having to keep up their cavalry forces, sent horses from Rome for this purpose : for, says my authority : 'There is no doubt of the fact that by these means our own breed of horses received this early cross in blood, as we have before seen.' Whatever was the nature of our own breed, it was thus early crossed with the Roman strain, in which we cannot doubt the Turkish horse was largely mingled, if its blood did not predominate over all others. During the period that Spain was under the servile yoke of Rome, the breed of horses must have had further infusion of mixed racial qualities by the crossing of the two breeds. The Arabian horse so-called, like the Turkish, had already found its way into Spanish territory. Consequently, we may conclude that, at or before the beginning of the Christian era, our horses were already mixed with the breed of four different nations. In fact, I suspect that they were, like horses in all countries, mixed in blood with nearly all the horses of the different nations upon the face of the earth.

Later there was yet another cross, when Athelstan got 'the running horses' from Germany, which we may conclude had their special merits, and were ultimately crossed with our own best breeds, with a view of improving or keeping up their excellence. This is the second cross from horses imported for the special purpose, so far as I have been able to ascertain. Strange to say, no record has been left as to the effect, good, bad, or indifferent; but from the fact that about this time, or shortly after, the Saxon king ordered that no horses should be sent abroad for sale or other purposes except for royal presents, we may fairly argue that a beneficial result was the effect of the cross; for, says Cecil in his useful little work on the subject: 'The English horses after this appear to have been prized on the Continent.' This is confirmatory of what Goldsmith had written long before.

The Spanish horse was known in England before the Arabian for improving our breed of race-horses. In 'Lewis's Topographical Dictionary,' under the head of 'Newmarket,' I find the following: 'A house called the King's House was originally built here by James I. for the purpose of enjoying the diversion of hunting; and the subsequent reputation of this town for horse-racing seems to have arisen from the spirit and swiftness of the Spanish horses, which, having been wrecked with the vessels of the Armada, were thrown ashore on the coast of Galloway and brought hither.' This would be about the year 1603.

We now reach a period when we can find something like an authentic account of the pedigrees and breeding of our race-horses, and of the stallions that were introduced into this country for the purpose of improving their blood. Messrs. Weatherby's 'General Stud-Book,' published in

ORIGIN OF EXISTING BREEDS.

1793, professes to give the pedigree, with few exceptions, of every animal of note that had appeared on the turf from 1743, and many of an earlier date, with some account of foreign horses from which the present breed of racers is derived. This is the oldest authenticated record of pedigrees of horses that I have seen. In this book mention is made of rather more than 700 brood-mares and most of their produce, about 350 other horses not included in the above list, and twenty stallions supposed to be of note, extending from about 1650 to the year 1793, or a little over a century.

The *Markham Arabian* was purchased by King James I., as I have related. *Place's White Turk* was imported later, in the time of Oliver Cromwell, as has also been mentioned. The horses, however, which by mating with the royal mares, and those already in the country, had most to do with begetting the progenitors of the animals now best accredited in the 'Stud-Book' were the *Byerly Turk*, the *Darley Arabian*, and the *Godolphin Arabian*. The royal mares, I should perhaps observe, were procured by Charles II., and the dam of *Dodsworth* was one of them. The *Byerly Turk*, it appears, was Captain Byerly's charger in Ireland. *Darley's Arabian* was the sire of *Aleppo*, foaled in 1711; *Almanzor*, in 1713; and two years later the celebrated *Childers*, or, as he was by some called, *Flying* or *Devonshire Childers*. From this it is probable that *Darley's Arabian* was imported about the year 1700. *Curwen's Bay Barb*, another noted sire, must have arrived much about the same time. For in 1711 he was the sire of *Brocklesby Betty*, who was thought to be superior as a racer to any horse, mare, or gelding of her time. *Curwen's Bay Barb* must have been very small, I should think, for he was the sire of *Mixbury* and *Tantivy*, 'both very high-formed galloways; the first of these being only 13 hands 2 in. high, and yet there were not more than two horses could beat him in his time, at light weights,' which does not say much for their merits or his.

Of *Flying Childers* report says 'that he was generally supposed to be the fleetest horse that was ever trained in this or any other country; that he ran a trial against *Almanzor* and the Duke of Rutland's *Brown Betty*, 9 st. 2 lb. each, over the Round Course at Newmarket, 3 miles 6 furlongs and 93 yards, in 6 minutes and 40 seconds, to perform which he must have moved at the rate of $82\frac{1}{2}$ feet in one second;' which we know is an absurdity on the face of it; and it is equally as much nonsense to say he ran a mile in a minute, as it is hard to believe that the Indian horses, *Antelope* and *King David*, have done the miracle in our own day. But that he must have been a very good horse in those times there can be no doubt.

Out of the twenty stallions mentioned in this work, I shall at this time only allude to one more, and that is the *Godolphin Arabian*, who was imported much later, and foaled in 1724. The breed of this horse (which in other places is called a Barb) is not known. His biographer, writing in 1793, says : ' In regard to his pedigree, from all that can be collected, none was brought over with him, as it was said, and generally believed, he was stolen.' Whatever pedigree, therefore, is claimed for him, must rest on the authority of some interested person, whose unscrupulous actions vouch little for the value of his word.

The *Godolphin Arabian* is, nevertheless, an animal concerning which a lively interest has always been felt. We hear he was a dark bay horse about 15 hands high, with a white off-heel behind. He was imported by one Cooke, at whose death he became the property of Lord Godolphin. Of his merits as a race-horse, like his pedigree, nothing is known, and by accident only he became fashionable as a sire. He was teazer to *Hobgoblin* for several years, and on the latter refusing to serve *Roxana* (a maiden mare), she was put to the Arabian, and from that cover produced *Lath*, a bay colt, the first horse that the *Godolphin Arabian* got. *Roxana* at the time (1731) was fourteen years old, and her partner then only seven. She bred *Round Head* by *Childers*, a sorrel colt, the next year, and in the year following, *Cade*, a bay colt, by the *Godolphin Arabian*. These were the only three foals the mare ever bred, dying the same year after parturition, the foal having being brought up by hand. The *Godolphin Arabian* was fifteen years old when he got *Regulus*, the sire of *Spilletta*, the dam of *Eclipse*. The following year he got *Babraham*, *Bajazet*, and *Blank*. From this account, taken from the 'Stud-Book' of 1793, it appears that most of his best stock was got before he was seventeen years old, and the last twelve or thirteen years of his life were not successful, as none are made special mention of, besides those I have enumerated. 'It would be superfluous,' it adds, 'to remark that he undoubtedly contributed more to the improvement of horses in this country than any stallion before or since his time.' How this is known I am at a loss to discover. He died at Gog Magog in 1753, in the twenty-ninth year of his age.

Many are the romantic stories that have been told of his love of a cat, and, as they widely differ, all cannot be true. In Messrs. Weatherby's 'Stud-Book' for 1803, we find the incident thus related, which is probably the correct version: 'The story of his playfellow, the black cat, must not be omitted here, especially as an erroneous account has got abroad, copied from the first Introduction to the present work. Instead of his grieving for the loss of the cat, she survived him, though but a short time : she sat upon him after he was dead in the building erected for him, and followed him to the place where he was buried under a gateway near the running-stable ; sat upon him there till he was buried, then went away, and never was seen again, till found dead in the hay-loft.'

After *Flying Childers* no remarkable horses were got by any of the sires imported into this country and crossed with our mares, for a number of years —nearly fifty. But then, most undoubtedly, the Arabian or Barb strain gave us, through the *Darley Arabian* and the *Godolphin Arabian*, the three best horses known up to that time, *Matchem*, *Herod* (sometimes called *King Herod*), and *Eclipse*. *Matchem*, bred in 1758, like *Flying Childers* and *Eclipse*, did not run till he was five years old. He ran at Newmarket over the Beacon Course (4 miles, 1 furlong, 123 yards), carrying 8 st. 7 lb., in 7 mins. and 20 secs., being about the same time *Flying Childers* is said to have made in his extraordinary feat over the Round Course.

Herod was very successful, and won many races at Newmarket. But, whilst running at York in 1766, he broke a blood-vessel, and was never so good after. He was put to the stud, and proved a good stallion. *Herod* was a descendant of the *Byerley Turk*, and the greatest proof of his excellent quality is found in the fact that he was the sire of

497 winners, whose winnings amounted to upwards of £200,000.

Eclipse was, by current report, the most remarkable horse of times gone by, of which we have any account: and the lapse of time has only served to fix his reputation. He was bred by Mr. Wildman in 1764, got by *Marske* out of *Spilletta*, foaled in 1749, got by *Regulus*, her dam (*Mother Western*) by Smith's *Son of Snake*—Lord D'Arcy's *Old Montagu*—*Hautboy*—*Brimmer* (*Marske* by *Squirt*, by Bartlet's *Childers*, by the *Darley Arabian*). He won eleven King's Plates, in ten of which he carried 12 st., and in the other 10 st.; the distance probably 4 miles each, or not much less. He was never beaten, and in one heat of a race distanced all his competitors. This feat, of which so much has been written, was accomplished to win a bet. His owner, Mr. O'Kelly, had injudiciously asserted he could place all the five horses running in the race; and this he did by placing '*Eclipse* first and the rest nowhere,' which in the language of the turf means that none of the rest would reach the distance-post before *Eclipse* had won. The feat was accomplished, the winner being more than 240 yards before the whole of his opponents. The fact that this seemingly extraordinary performance had been done by others may have made Mr. O'Kelly desirous of imitating a feat he could not surpass. In one instance the task was achieved at Guildford, where, curious to relate, the £50 race won by Mr. Black's bay horse *Hazard* had the same number of runners, all of whom he distanced like *Eclipse*, and, by a singular coincidence, in the second heat. The total winnings credited to *Eclipse* were £25,000. This must, I imagine, have included bets, as it is in another place recorded with greater probability that he won but £2,000 in stakes, and retired early from the turf to the stud, where he was considered eminently successful, being the sire of 134 winners, and the amount of their winnings in the aggregate is set down as no less than £160,000. So, in those early days, stakes were run for of much greater value than is generally supposed, many being matches for 1,000 guineas each and upwards.

Marske, as the sire of *Eclipse*, deserves to be mentioned, and cannot be noticed in a more fitting place than immediately after his extraordinary son. He was by *Squirt* out of *Black Legs Mare*, and is traced back to *Hutton's Grey Barb*, and in the eighth generation to *Bustler*; but here all traces of his blood are lost, and it is not recorded how his dam was bred. The pedigree of *High Flyer* afforded a similar instance to that of *Eclipse*, for on the side of his dam he can be traced to a royal mare; but in the paternal line his genealogy terminates with *Bustler*, which horse was the son of the *Helmsley Turk*. In this fact we see a strange coincidence, as both horses are supposed to have been the best of their day, as good on the turf as they were at the stud.

Here our inquiry as to the origin of existing breeds may well conclude. To these three sires, *Herod, Eclipse,* and *Matchem,* all existing thoroughbred strains of most value may be traced. We find their blood in the three horses which are undoubtedly the best stallions of our time—*Touchstone,* foaled 1831, *Voltigeur,* foaled 1847, and *Stockwell,* foaled 1849. In these three grand sires I consider that every requisite in respect to 'blood' that the breeder requires is found; and in selecting his stock he has but to trace back to one or the other of them to be assured that all essentials in this respect are secured.

CHAPTER VI.

REAL VALUE OF THE ARAB STRAIN.

Alleged superiority of the Arab—Upton's 'Newmarket and Arabia'—Advantage of the cross apparent only in later generations—The *Godolphin Arabian* as a sire; not equal to present stallions—Early Arabs not successful on the course or at the stud—Instances of improvement as the cross recedes: the *Bald Galloway;* 18 stone to victory; other examples—Instances to the contrary from earlier horses—Failure of the cross since *Eclipse's* time—Authorities on my side: Buffon, Goldsmith, 'Nimrod,' Admiral Rous, Cecil, the *Sporting Life*—Covering fees of noted Arabs—A contrast—How much 'temper' may be due to the strain—Protest against revival of the practice —Failure of Hampton Court experiments.

On the subject of the breed of the English thoroughbred, something requires to be said on the assumed and real value of the introduction of the Arab strain. I have already observed that through this strain we undoubtedly obtained *Matchem, Herod,* and *Eclipse,* the progenitors of the best stallions of our time. But the believers in the 'Children of the Desert' are not contented with this admission. They claim for their *protégés* all excellences under the sun, and would have it—first, that the Arabian horse excels all others, and secondly, that in the present day our English thoroughbred would be improved by further crossing with Eastern blood. This is an assumption that I cannot permit to pass unchallenged. If there be truth in it, it ought to be looked to; if it be wrong, the theory should be at once refuted.

The latest treatise on breeding which I have seen is Captain Roger D. Upton's 'Newmarket and Arabia.' It does not, however, teach us much about breeding. Its chief object is a comparison between the Arabian and the English horse. For this purpose it gives an elaborate account of the Eastern horse, his habitat, pedigree, beauty, and excellence; and, indeed, avers that the Arab is the only thoroughbred horse, and that all our English horses are a degenerate or half-bred race. It is, I have no doubt, a work of much labour and erudition, though of little importance, except to those who care to trace the horse of the present day back to the earliest times on record.

It is, however, impossible to follow Captain Upton in his fanciful account of the wanderings of the horse after leaving Mount Ararat, until in imagination he locates it in Arabia, in a sort of magic circle untouched and untouchable. And on a basis with this conjecture must be placed the different theories set forth to show that the pure Arab blood, without admixture of any kind whatever, is incomparably superior to that of our horses.

The truth is that we do not know if the Arab horse, as imported early in the last century, was really so superior to the best class of horses then existing in the country as is generally believed. That he is to-day very inferior to our own horses, I shall have no difficulty in showing, when we come to the point. But as to his alleged superiority even at the earlier date, there is no real evidence. Certainly we have no record of a complete ascendancy of imported Arabians on the race-course, whilst the improvement of our breed by their introduction to the stud was, as I shall presently show, generally more apparent in the third and fourth than in the earlier generations. The value of the produce was, therefore, largely due to the inherent quality of the native element. The cross with the Arabian was an extremely fortunate one, as apparently just giving certain qualities which our own horses lacked at that distant period. But for all practical or useful purposes in this country, the Arab horse might to-day be as defunct as the dodo, or the huge mastodon of North America.

I have not advanced so much without being prepared to back my argument with facts. I admit that the introduction of the Arab strain achieved a great and happy result at the time; but I am also prepared to show that the best result was only obtained in later generations; the improvement continuing up to our own time is, therefore, primarily due to our climate and the qualities of our native horses; and further, that the Arab horse probably never was, and certainly is not now, as good as our own.

The most famous of these imported Eastern horses was undoubtedly the *Godolphin Arabian*, himself of a nameless race, of whose successes at the stud I have given a full account from a trustworthy source in my previous description of him. But even this horse, as a stallion, has been spoken of in a way he hardly deserves perhaps; for though undoubtedly he was a 'Triton among the minnows,' all that is positively asserted of him is that he was the sire of six good horses already named. It is added 'of many others,' but this, after all, is but a vague term, little to be relied on, to prove he was anything more than a good, fair stallion. We may, indeed, conclude that he was no more nor less, if we remember the fact that he was twenty-one years at the stud, during which time he would be the sire of six or seven hundred horses, if he only got thirty foals a year. And from the expectation he raised early in life by his sons *Lath* and *Cade*, he would probably have had most of the best mares. To the following stallions he certainly will not bear a favourable comparison: *Venison, Stockwell, Touchstone, Orlando,* and a dozen others in days gone by; nor would he to such horses as *Galopin, Sterling, Isonomy,* or *Hermit,* whose covering fee is 250 guineas each mare; nor indeed would he to many of the second or third class horses of the present day.

The *Godolphin Arabian* is often spoken of as the last of the Arabians or Barbs imported into this country. However, that is not the case. Several have been since imported, even up to the present time; but what can be certainly said is, that he was the last of special note. We have seen what he did, and now may proceed to consider the merits of the other celebrities which preceded *Eclipse*. In the 'Stud-Book' before referred to will be found the names of some twenty stallions, with their progeny attached. These include the *Byerly Turk, Place's White Turk, Dodsworth, Greyhound, Curwen's Bay Barb,* the *Toulouse Barb,* the *Darley* and the *Godolphin Arabians*. Besides these there are several pure bred Barbs which were foaled in this country, and therefore had all the advantages of acclimatization. Yet, strange to say, not one of them is mentioned as having been a good runner. From this circumstance it is more than probable that they were not first-class even in those early days of racing. We may be sure, had they done anything of note, it would have been duly trumpeted forth to add to their already overrated fame. Even as stallions they were by no means uniformly successful; and therefore there is little to be said to uphold the belief that no other horses could equal them.

It is only when we come to horses removed one cross or more from the Arabian, Turk, or Barb, on one side or the other, that we discover really satisfactory performers. I will give first an account of a celebrated mare, *Bald Charlotte*, retaining the quaint language of the original record:

'*Bald Charlotte* was a mare of shape and beauty, also of size, and had a very grand share of both speed and goodness. She was bred by Captain Appleyard, of Yorkshire; her sire was *Old Royal; Charlotte's* dam was a daughter of the *Bethel Castaway*, her grand-dam was a gray mare of Captain Appleyard's father, got by *Brimmer*. *Bald Charlotte* at five years old, viz., in 1726, beat twenty-three mares for the King's 100 guineas at Black Hambleton, in Yorkshire. The same year she beat seven in the Contribution October Stakes at Newmarket; and at that place also, on the 15th of April, 1726, she won the King's 100 guineas for five-year-old mares, 10 st., one heat. On the 18th, ditto, 18 st., she beat Mr. Ashby's *Swinger*, 17 st. 7 lb., 4 miles 300 yards. The same year, at Winchester, she won the King's 100 guineas for six-year-olds, weight 12 st. On the 20th of April, 1729, she beat, 9 st. 2 lb., Sir R. Fagg's *Fanny*. She has been the dam of several foals, and became the property of His Grace Charles, Duke of Somerset.' (Published March 10th, 1756.)

This description of the performances of a celebrated mare appears to be rather extraordinary. From her appearance in a print that I have seen of her, in an excellent state of preservation, she seems to have had little of the Eastern blood in her. She has a Roman sort of nose, and stands rather high on her legs, and very light in her back ribs. In beating no less than twenty-three mares for the King's Plate at Black Hambleton (a number I never remember to have seen or heard of running for such a race before), and carrying 18 st. to victory, she performed feats which are curious events in the life of any race-horse, and well worthy of record. The weight that she carried is quite beyond modern parallel. I once saw 14 st. 5 lb. carried by the five-year-old *Chandos* three-quarters of a mile at Stockbridge, when, ridden in a 3 lb. saddle by his noble owner, the then Lord Aylesford, dressed in trousers, he gallantly defeated 'a small field.' This was in 1875, and was thought much of at the time. It was also said of this wonderful mare that she bred several good horses, and was great-grand-dam of *Coxcomb* and *Dorimant*. She was, anyway, an extraordinarily good runner and a successful brood-mare—but she was not an Arab. Indeed, the nearest strain we find in her is in her great grandsire *Brimmer*, himself one remove from a Turk.

Amongst other good animals, I find the following:

Bay Bolton, by *Grey Hautboy*, was a good runner and proved himself an excellent stallion.

Little Driver, by *Beaver's Driver*, dam by *Childers*, was a good runner, having won upwards of thirty £50 Plates.

Fox, by *Clumsey*, bred in 1714, out of *Bay Peg*, by the *Leeds Arabian*, was the sire of Captain Appleyard's *Conqueror*, the best gelding that ever ran at Newmarket, of *Merry Andrew* and *Goliah* (good horses). He also got the dam of *Snap*.

Partner, by *Jigg* out of sister to *Mixbury*, by *Regulus*, was the sire of *Sedbury*, *Tartar*, *Cato*, *Traveller*, and ten others mentioned by name in the 'Stud-Book,' besides many more that are not.

Sloe was by *Crab*, dam by *Childers*. He was never beaten, having won five Royal Plates and two other prizes in one year, and he was the sire of *Sweeper*. I shall refer to him again as a stallion later on.

Starling, by *Bay Bolton*, his dam by a son of the *Brownlow Turk*, was an excellent runner, and the sire of *Skim*, *Ancaster*, *Starling*, *Torrismond*, *Teazer*, *Moro*, *Jason*, and the grand-dam of *Soldier*.

Tartar, by *Partner* out of *Meliora*, by *Fox*, was an excellent racer, and not less esteemed as a stallion, being the sire of *King Herod*, and of Mr. O'Kelly's mare, dam of *Mercury*, *Volunteer*, and many other good horses.

Traveller, by *Partner* out of a daughter of *Almanzor*, won several Plates and got many good winners.

These examples of horses of more than fair merit are all removed some distance from the original Eastern cross. As we come nearer the Arab strain, in fact, we find the performance less satisfactory, many being failures on the race-course, and some of them equally bad at the stud.

Almanzor, by *Darley's Arabian*, was a public stallion, and covered a great number of mares; but though he was an extraordinarily fine horse, and very well bred, he got very bad horses. Of his racing merits nothing has been said, so I suppose he possessed none.

The *Bald Galloway*, by *St. Victor's Barb*, got *Cartouch* (a capital galloway at five years, but trained off), and other middling horses, mentioned by name, as well as some galloways.

Hartley's Blind Horse, by *Holderness Turk*, was the sire of the large *Hartley* mare, dam of *Babraham*, and of the great-grand-dam of *Priestess Espersykes*.

Lastly, I may mention *Jigg*, by the *Byerly Turk*, who got *Partner*, a capital horse, *Shock*, and *Saucebox*, middling horses. He was a common country stallion in Lincolnshire till *Partner* was six years old.

These facts are taken from the second part of the 'Stud-Book,' which gives the pedigrees and the performances on the turf and at the stud of over two hundred and fifty horses. And from them I think it will be plainly seen that, however advantageous to our breed of horses was crossing them with Turks, Barbs, or Arabians in the first instance, the advantage was in the cross itself, in supplying some deficiency in our own strain, and not in the actual superiority of the Eastern horses. In fact, the advantage, whatever it was, only appeared in a decided manner in the second and succeeding generations; and, in short, the further we recede from the original strain, the better our horses became up to the time of *Eclipse*. Since his time, any experiment in crossing back with Oriental stock has proved a most decided failure. How poorly, indeed, do the earlier specimens, the galloways of 13 hands 2 inches to 14 hands high, contrast with the thoroughbreds of the present day, many of them standing 16 hands and over! Horses, too, well proportioned, and able to carry 16 stone to hounds, and run any distance, such as *Stockwell*, *Springfield*, *New Holland*, *Fiddler*, and *Bendigo*.

Whether the Arabian or Barb once was, as some think he was, superior to our own breed as a race-horse, or whether, as many good judges have unmistakably pronounced him to be, he was inferior, there can be no doubt of his inferiority to-day. As to his alleged superiority when introduced here, we are not without eminent authority to rebut the assumption. Buffon, in the heyday of the *Godolphin Arabian*, must have been in the prime of life, and was probably employed in writing his celebrated 'Natural History,' in which he says, ' No horse can equal our own, either in point of swiftness or strength.' In this opinion he is strictly corroborated by Goldsmith, who wrote about the year 1760, at the very time this fierce contest was so hotly raging regarding the degeneracy of our own horses or their want of goodness, and the superiority of the Barb, and other foreign horses which were introduced for the improvement of our own breed. After describing the horse most minutely, with the different purposes for which he is used, and his capabilities, Goldsmith sums up the whole in these few but significant words : ' I have hitherto omitted making mention of one particular breed, more excellent than any that either the ancients or moderns have produced ; and that is our own.'

Now I think I may fairly ask who is to contravene the evidence of two such painstaking and able writers, one actually living at the date ? And if they are right, which I think none will deny, what is to become of the tenets of later chroniclers, who are always asserting, in a desponding mood, that our horses at the time were good for nothing, or at any rate not better than hacks of the present day, if even as good, and that all imported horses were nonpareils? It is not a conclusion which I can reach. Rather I am inclined to think that our horses at that time were much better than by many they are supposed to have been.

The truth is that the superiority of our horses has been distinctly proved over and over again, both in speed and endurance, whenever the two breeds have met, whether on our own turf or on the sandy desert. Nor do I want support from the most eminent authorities in the view I take. For, in the quotation elsewhere given from 'Nimrod' as to the spread of English blood throughout the world, not a single instance is adduced of any horse deriving its blood from the Arab, but all from the English. 'Indeed,' he concludes, ' the Arab horse is not even mentioned in all these different countries as having contributed to the breed of any.' This was written fifty years ago. And since then, I need scarcely say, the pages of the 'Stud-Book' prove to what extent the world is indebted to the English horse, and the English horse alone, for the improvement of the breed.

I can scarcely quote a better authority than the late Admiral Rous, who, in an important passage that also will be found in another page, under the head of 'The Improvement of the Modern Horse,' fully confirms my opinion that, highly prized as Eastern blood was 150 years ago, and much as our present breed is indebted to the cross, our horses were then, through earlier crosses, already much better than formerly, and, if ancient writers are to be believed, were better than the Arabians themselves. And what is still more important, his opinion coincides with mine that the Arab horses of that date were in reality much about the same as they are to-day; that is, little above mediocrity in size, speed, and stamina. These are his words in the passage referred to: 'They are no better now than they were 200 years ago.' And I may add we know what they are now; simply that the best animal amongst them would not win a good selling plate in the present day. I refer to the passage alluded to, which is given later, for a further exposition of the gallant Admiral's view.

Another admirable authority, 'Cecil,' has the following confirmation of my opinion:

'The idea' (he says) 'of introducing Arabian blood into the stud for racing purposes would, I feel assured, be treated by every practical and experienced breeder with contempt; and I only venture to mention the subject in order to point out some of the most ostensible reasons why the attempt would be followed with disappointment. In the first place, the breed of race-horses in this kingdom is far superior to the Arabians or other foreign horses brought to England at the present period. The casuist may remark that our own blood was originally derived from these sources, the reply to which is obvious. They have taken kindly to the soil, the climate, and the treatment; hence the present breed of horses is superior.'

Lastly, I shall quote from an eminent contemporary writer, the passage I allude to being taken from an able article on breeding in the columns of the *Sporting Life*:

'An Arab' (says the writer) 'is invariably a bad hack on the road, and his legs will not stand the same knocking about as those belonging to his English relative. Take the Arab as he is, there is very little about him that can compare with our own horses. He has nothing like the pace; he is always a poor performer over a country as compared to an English hunter; and if his endurance is unquestionable, he has been beaten at that also by moderate English race-horses, and in some instances by hacks.' In short, the Arabian is as seldom seen in the hunting-field as on the race-course, being fit for neither one place nor the other, and worse as a roadster.

It will not be uninteresting here just to inquire what was the value set upon the Arab strain at the time it was in most request. *Babraham*, in point of size—16 hands high—a monster to most horses at that date, covered at a fee of two guineas, and two shillings for the servant. This seems to have been about the average price; for although *Sloe*, who was never beaten, covered at three guineas and half-a-crown for the groom, we find the one following *Babraham* in the list of stallions described as 'a fine strong bay horse, 14 hands and 3 inches high, covering at fifteen shillings a leap and trial, and one shilling the groom.' I shall mention but one more, that is *Trifle:* 'He was got by old *Fox*, son of *Clumsey*, son of *Hautboy*, son of the *White D'arcy Turk*. He has won several £50 prizes, and will cover at one guinea a mare and one shilling the man.' Let us for a moment compare these prices with those of our own sires at the present day, covering at 200 guineas, and though not all publicly advertized, some at a much higher figure, with the subscription lists generally full. And these, be it remembered, were their best horses. For, in writing of *Babraham* and his merits, I am reminded of an anecdote related by Sir Francis Hasting Doyle, who tells us that Samuel Johnson interested himself in *Atlas*, by *Babraham*. 'Johnson,' says Sir Francis, 'went his way muttering to himself: "Of all the possessions of the Duke of Devonshire I covet *Atlas* the most."' Holcroft, in his memoirs, says the trainers at Newmarket thought him the best horse that had run since *Flying Childers*.

I have not yet mentioned another objection to

Eastern blood. I lately had a conversation with a gentleman in the army who was in India during the Mutiny, at Cawnpore and other places, and he told me that 'most of the Arabians or horses of Eastern blood are bad tempered, and none but the black attendants dare approach them, for they have been known to seize their owner and kneel on him when down, and many have escaped only with their lives.' Who knows, then, but what we are more indebted to the Eastern blood than to any other breed for the savage propensities of our horses, which make many an otherwise valuable horse useless at the stud?—that is, in my view of the matter.

Before abandoning the subject, I may allude to the growing disposition to reintroduce the Arabians by giving races in which they alone can compete, having for its ultimate aim, I suppose, the crossing of them with our mares. If so, I hope it will be nipped in the bud and condemned by all right-thinking men, so that we may not see another failure such as our fathers before us have seen and regretted. The five or six Arabian stallions lately sent by the Sultan of Muscat to the Queen are very likely to be put to this purpose, and tried at Hampton Court, with about as much probable success as was witnessed in the time of William IV., when no purchaser could be found that would give £40 each for the best bred Arabian yearlings, or for horses out of Arabian mares, and the attempt to sell them had to be abandoned. I have, perhaps, considered the subject more exhaustively than may be thought necessary by some; but if books, not pages, were written, which would only convince the breeders of thoroughbred stock of the utter uselessness of the Arabians for their purpose, the labour would not be lost. In my next chapter I shall consider the remarks of Captain Upton on the Arabian horse and his breeding.

CHAPTER VII.

IMPROVEMENT OF THE MODERN HORSE.

Existing disbelief—One cause of this : ' short courses '—Difficulties of the breeders of judging—' Gameness ' of our present horses—Admiral Rous's testimony—What horses really did a hundred years ago—' Sweeping the board ' in those days ; a ten-guinea prize—Performances of *Eclipse* examined—*Eclipse* and *Touchstone* contrasted on the turf and at the stud—The lesson therefrom—Advantageous effects of climate ; the Arabian ' a curious exception '—Our improved fat and lean stock—My Russian pig ; ' Early Porcine Type '—Spread and appreciation of the English horse in other countries ; ' Stonehenge ' thereon ; what ' Weatherby's List ' says ; ' Nimrod's ' testimony—Progress in the colonies—New Zealand.

ONE of the assertions very easily made, and very readily and widely believed by the inconsiderate on proof that cannot even be qualified as ' slender,' is that the English horse of to-day is not the equal, in certain respects, of his predecessors. It is a question that I may treat, for it is well to know, whilst we are discussing the subject of breeding, whether the process followed has resulted in improvement or failure. I may say at the outset that the supposed degeneracy of our thoroughbred horses is altogether a mistake. It is the manner and the purpose for which they are used that gives them the delusive appearance of non-stayers, in comparison with horses of an earlier date, when long races and heavy weights were the rule, and not, as now, the exception. If horses of the present day were trained for long races and ran but a few times a year, this important fact would be made manifest. Five or six furlong races are so numerous now that few trainers care to prepare their horses for any other distance, or at most only a little beyond it. In this they have the tacit sanction of their employers, for there is scarcely one nowadays but likes to see his horses big when brought to the post; a state totally incapacitating them for running a long distance successfully.

This is a much more important matter than one might take it to be at first sight. Breeders have many difficulties to contend against; and this is not, by any means, one of the least of them. Public performance becomes no guide whatever to real merit in that one point, 'gameness.' It is only necessary, in order to make this clear, to notice the sort of horses we get running in our day in short courses. We see *Charon, Jester, Hampton*, and indeed a host of others that could be named, running in selling races, and mostly over short courses; and if any of these horses had retired from the turf at the end of their two-year-old career, from accident or otherwise (as many horses do), they would have been stamped as non-stayers, and have been looked on at the stud as virtually only fit to get short runners; whereas, truly, they were all good game horses, and only wanted a distance of ground to show their intrinsic merit in its true form. This is really a difficulty; for how can we be sure, in the present day, that horses are not passed by for stud purposes as 'non-stayers' which in truth only lacked the opportunity to show their capabilities?

As to the real improvement in the modern horse, I will proceed to quote the passage, alluded to in a former chapter, from the work by Admiral Rous:

'A very ridiculous notion exists,' says the gallant handicapper, 'that because our ancestors were fond of matching their horses four, six, and eight miles, and their great prizes were never less than four miles for aged horses, that the English race-horses of 1700 had more powers of endurance, and were better adapted to run long distances under heavy weights than the horses of the present day; and there is another popular notion, that our horses cannot now stay four miles. From 1600 to 1740 most of the matches at Newmarket were above four miles. The six-mile post in my time stood about 200 yards from the present railroad station, six-mile bottom, and the eight-mile post was due south from the station on the rising ground; but the cruelty of the distance, and interest of the horse-owners, shortened the course in corresponding ration with the civilization of the country. Two jades may run as fine a race for eight miles as for half a mile—it is no proof of endurance. You may match any animals for what distance you please, but it is no proof of great capacity. We have no reason to suppose that the pure Arabian of the desert has degenerated; his pedigree is as well kept, his admirers in the East are as numerous, and his value in that market has not been depreciated. In 1700 the first cross from these horses were the heroes of the turf. Look at the portraits of *Flying Childers, Lath, Regulus*, and other celebrated horses, including the *Godolphin Arabian*. If the artists were correct in their delineations, they had no appearance of race-horses; they, of course, were good enough to gallop away from the miserable English garrons of that era, as a good Arab or a Barbary horse, like *Vengeance*, would run away from a common hackney in the present day. Amongst the blind, a one-eyed man is a king. My belief is that the present English race-horse is as much superior to the race-horse of 1750 as he excelled the first cross from Arabs and Barbs with English mares, and, again, as they surpassed the old English racing hack of 1650. The form of *Flying Childers* might win now a £30 Plate, winner to be sold for £40; *High Flyer* and *Eclipse* might pull through in a £50 Plate, winner to be sold for £200. This may be a strong opinion; it is founded on the fact that, whereas 150 years ago the Eastern horses and their first cross were the best and fastest in England, at this day a second-class race-horse can give five stone to the best Arabian or Barb, and beat him over any course, from one to twenty miles. I presume, therefore, that the superiority of the English horse has improved in that ratio above the original stock.'

Merely noticing in passing that the gallant Admiral fully confirms in this passage my testimony as to the quality of the Arab as compared with the English thoroughbred, especially as regards his running to-day, when 'one of our second-class horses could give five stone' to the best of the lot, I will supplement the information he gives by adding an exact account of what the race-horse absolutely did in the year 1750, of which we have accurate information in the 'Racing Calendar' for that season.

There were then 405 horses running, belonging to no less than 241 owners; a circumstance from which we may learn that none of the studs could compare in magnitude with the size of many at the present day; indeed, few could exceed five or six public runners. Instead of being capable of performing feats of endurance, we find that two horses only ran as many as five races each, in that year. These performances afford a curious contrast. *Driver* won all his races, which were run for in heats, or in other words ran eleven times; the last of them reading like a race for galloways at Newmarket: '£50 Plate, 14 hands, to carry 8 st.; 7 lb. to give or take.' The other, Mr. Greville's *Noble,* was beaten in all his races but one at Salisbury, the winning of which I should think was an accident, for I find his single opponent was distanced. Mr. Humphrey Sturt, an ancestor of the present Lord Alington, hailing from Dorsetshire, apparently was a shrewd and intellectual person, who raced, and had a well-named horse called *Nothing,* which was distanced for the £50 Plate at Aylesbury, and ignobly retired from the turf; or, at any rate, took leave of it for the remainder of the season—a decision his lordship would most certainly have acquiesced in had he lived in those days, or had he such an animal running in these.

We must remember, too, that all races were not in those days run in heats, for I see that at Black Hambleton four days' races consisted of four races, not one of them in heats. Nor, and what is perhaps equally important, were these rich stakes all of the full value of fifty guineas, or about it. For we see Mr. Sparrow's chestnut gelding *Cripple* was the winner of the annual ten guineas at Barham Down, Canterbury, beating all his opponents, and triumphantly retiring with 'all his blushing honours thick upon him.' Long before this date we read of races to be run for of much less value. In 1711 'a plate of six guineas value, three heats, by any horse, mare or gelding that hath not won above the value of £5: the winning horse to be sold for £10, to carry 10 st. weight if 14 hands high: if above, or under, to carry or be allowed weight for inches, and to be entered on Friday the 5th, at the Swan Inn, Coleshill, by six in the evening. Also a plate of less value, to be run for by asses.' Out of the remaining 402 horses, 240 of them followed the good example of the *Cripple* and only raced once each in public; but through not all having the same incomparable powers, many had not the same good fortune of conquering every competitor and retiring without an equal.

And when we come to really good horses, there are other matters to be taken into consideration before we can gauge the relative merits of horses now and in *Eclipse's* time. Horses may, through accident, after winning, leave the turf; others may win a race and die, and so may be said never to have been beaten, and thus gain a reputation their merits do not entitle them to. I mention this to show that after all *Eclipse*, *High Flyer*, and *Matchem* may not have been the equine wonders of all time that

IMPROVEMENT OF THE MODERN HORSE. 87

people generally repute them to have been. In those days each horse ran but few times, and generally against small numbers, and those most likely of an inferior class. Indeed, everything goes to show that such was the case; and as we can only find about two or three good horses annually in 2,000, how is it likely they should have found so many of the same quality in about 600, the number running in those days?

Again, if we compare what *Eclipse* did as a race-horse and at the stud, not with our best horses now, but with them half a century back, and take *Touchstone* as an example, we shall find the result very much in favour of the latter. He ran earlier, oftener, and won as much in one or two stakes as *Eclipse* did in all his races put together. He was longer on the turf, and died at a more advanced age. His stock won 771 races, of the collective value of over £228,000; and to win a race then must have been, as now, much more difficult than it was in the days that preceded, namely, about the time of *Eclipse*, when, according to an account taken from the Duke of Tuscany's travels in 1669, 'hacks were as good as race-horses.' 'For,' he says, 'the English horses, being accustomed to run, can keep up with the racers without difficulty.' But even in those days *Eclipse's* stock of 134 winners and the amount of £160,000 does not show a favourable contrast to *Touchstone;*

and much the same may be said of the best horses that succeeded him.

I do not, for a moment, make the comparison for the purpose of tarnishing the well-earned fame of great horses. I do not wish to make too much of the fact, set forth on the irrefragable testimony of the 'Racing Calendar' of 1750, as to the turf paragons of the period retiring amidst general acclamation after winning one ten-guinea stake. But I do say that if these things were considered we should hear no more of the cuckoo-cry about 'the degeneracy of our horses.' And more, that these truths are infallible guides to the proper estimation of the merits, or demerits, of different animals at different periods of time. I have no doubt myself that we have seen as many and as good, or better horses, within the present century; and more, that there are a greater number running at the present day than were ever known at any other period of turf history.

I do not wish to lead to the conclusion that, having established the improvement in the horse as a matter of fact, I attribute it entirely to greater knowledge or to greater care in his treatment. That we have learned not a little, and that we have better means of preserving his health, is true. But it does not do to ignore the effect upon successive generations of the influences of climate, in causing improvement by a simple natural process.

We see this in other animals besides the horse, and it is no more easy to account for it than for the process of natural selection.

If we are to believe Captain Upton and other enthusiasts, there is one animal, indeed, that is above such influences, whether occult or demonstrable. According to them the Arabian horse alone, of all animals in creation under the control of man, is incapable of improvement, either by judicious crossing with his own pure blood or with a better. I will not stay to inquire whether, just as our own race-horse has been improved by breeding from different strains, the Arabian might not possibly be improved by a cross with our stallion. I am not concerned with the improvement of the Arabian. That is a matter I leave to the ability of more ardent admirers of his race than myself. The question I have to treat of is the improvement of our own horses.

The effect of the influence of the climate and soil of this country, in furthering the improvement of cattle of all sorts, is not a matter that should be slighted in a work on breeding. For we are indebted to it not only for the superiority of our breed of horses, but also for the surpassing excellence of every description of stock that we raise, either for our own consumption or for the improvement of our own or foreign cattle in the most distant regions of the earth. By this essential and primary assistance, combined with skill and indefatigable perseverance, we have excelled all other nations in the production of cattle, sheep, and pigs, to an extent that to be appreciated must be seen in the various exhibitions of fat and lean stock in the leading agricultural towns, and at the Christmas shows in London and Birmingham.

As to the influence of climate, a matter which I shall deal with at length when I come to describe the condition which should guide the breeder in the choice of locality and soil, I am tempted here to mention my own experience with 'my Russian pig.' Shortly after the Crimean War, Mr. Wolfe kindly made me a present of a young Russian pig. This lanky creature, which had been brought over by an officer on his return from the Crimea, was in shape more like a greyhound than a pig, and nearly as swift, with bristles longer and stiffer than I ever saw before on any of its species. I kept it for some time; but as its appetite, though not nice, was voracious, and as it had no tendency whatever to fatten, I gave it away, and so lost sight of it. Surely, if this were a fair specimen of the 'early porcine type,' we may judge, by comparison of it with what we can produce to-day, of the extent to which climate and treatment may improve any kind of animal. Fancy such a creature at one of our exhibitions, standing in a pen side by side with our best-bred Berkshire pig, or the large white

Yorkshire kind, or other and smaller breeds that are now dispersed throughout the country! Then, contrast a good fat Southdown sheep with its original stock, the moufflon, and it will be manifest what selection has done for us in effecting the incomparable excellence of the sheep at home and in our Colonies. So, with our horses, there can be no doubt that they are much improved, and better adapted for the various purposes for which they are used, than they were one, two, or three centuries ago.

As to the world-wide tacit acknowledgment of this fact, shown in the gradual and increasing spread of the English horse in other countries, I may, in concluding my observations on this part of my subject, quote facts from well-known authorities. I find that 'Stonehenge,' writing about the year 1830, says that up to about then we had exported to America no less than 392 pedigree horses. But, from the pages of Messrs. Weatherby, I take a still more startling record of the number of horses that have been exported, not only to America, but to nearly every place upon the face of the earth, to wit — British North America, South America, the United States, Australia, Austria, Hungary, Belgium, South Africa, Germany, India, Italy, New Zealand, Poland, Portugal, Roumania, Russia, Spain, and Sweden, in about three years. The numbers give a total of 924 mares and horses exported in this period (practically within this period, that is to say, for in the number are included a few exported before 1877, and a few in 1881, up to June, the date of publication). Of these, 159 were sent to France, and 329 to Germany.

The following extract from 'Nimrod' is much to the same purpose. 'After the example of Englishmen,' he says, 'racing is making considerable progress in various parts of the world. In the East Indies there are regular meetings, and there is also a Royal Jockey Club. In the United States, the breeding and running of horses are advancing with rapid strides. In Germany we find there are many regular places of sport. His Serene Highness the Duke of Holstein-Augustenburg and his brother Prince Frederick have each a large stud of horses, from blood imported from England; and many other German sportsmen have studs under the care of English stud-grooms. Prince Butera's breeding-stud on the southern coast of Sicily is the largest in those parts. It was founded by a son of *Haphazard* from a few English mares, and His Highness is one of the chief supporters of Neapolitan horse-racing. In Sweden is some of our best blood, and Count Woronzow and others have taken some good blood to Russia; and as to racing in Tasmania, several well-bred English horses are the best of the cattle at Hobart. Racing in Germany is considerably on the increase, particularly in Hamburg and Berlin, in Mecklenburg and Holstein, as well as in the whole of Germany and Prussia, showing that no expense is spared in purchasing the English blood.' Again he says, ' But it is in the New World, America, that racing, and the consequent improvement of horses, is making the most rapid progress, having twenty-nine thoroughbred English horses propagating their stock throughout the various stations.' After giving the names of the stations, and a list of the horses referred to, which I need not repeat, he continues: ' And to these are to be added *Glencoe*, and, alas ! *Priam*, at the extraordinary cost of 3,500 guineas.'

This was written some time ago, and the arguments in favour of the growing demand for English stock are only strengthened by later developments. 'Nimrod,' it will be seen, refers to the progress made in Tasmania, and as further illustrating the appreciation in which the modern English horse is held in the colonies, I may mention that, in the seventh volume of the ' Stud-Book ' for New Zealand, published in 1881, I find great strides have been made there in breeding thoroughbred stock. It includes nearly 1,000 mares and their produce. Many mares have been imported, and most of them are descendants of imported horses from this country. In the obituary of stallions is a list of eighteen, fourteen of which are either by imported stallions or out of imported mares; three home-bred, and one Arab. Under the head of 'Covering Stallions,' I find no less than sixty-one horses that are either imported, or from imported horses, or more than half the number that is advertised to cover in our 'Book Calendar.' Amongst the horses imported from this country I find *Musket* (perhaps nearly as good as any stallion we have here at the present moment), *Leolinus, Feve,* by *Lord Clifden,* and the following that won me several races before I parted with them: *Castle Hill, Traducer,* and *Mail Train.* So they have some well-bred horses in the country, and, like all sensible people, have but little to do with the Arab blood. Amongst the mares are *Pulchra, Dundee's Katie, Crinoline, Forget-me-not,* and *Sissie,* than which we have few if any better bred mares in our own 'Stud-book.' There is, in short, scarcely a place on the face of the earth where races are held in which the successful competitors do not trace their purity of blood to our own horses. And I confess· I cannot see in what direction any but crotchet-mongers can trace a sign of the degeneracy, and not of the improvement, of the modern horse.

CHAPTER VIII.

CLIMATE AND OTHER INFLUENCES.

Climate and soil must be suitable—Abundance of eligible farms—Effect of our climate on horses in size and shape—Undesirable localities—Goldsmith on climatic effect on dogs; on improvement in cattle—The wild herd at Chillingham—Progressive advance in breeding—Increased weight and value of sheep—Weight of a modern ox—Contrast of past and present times.

Other influences—Further requirements in selecting farm—Abundant pasturage—Sufficient room—Effect of exercise on action—Essential need of it—Example from greyhounds.

WHEN, in the first chapter, I expressed my conviction that farmers, large and small alike, could successfully add to their ordinary employment the breeding of horses, I named one necessary condition of this success—that the climate and soil should be suitable for the purpose. Lest, however, this proviso should appear at all discouraging, I may add that, as a rule, farming land throughout the United Kingdom is suitable; the exception being in the condition or surroundings of certain localities which I shall presently describe.

Climate, undoubtedly, has much to do with the

improvement of all animals, so much so that the process in this respect is one of the great mysteries of Nature. Even in our own little island, small as it is, horses could not be bred so good in one part as in another. No one, for example, would think of trying to breed thoroughbred stock on the Welsh or Scotch mountains; or, for the matter of that, indeed, even on any of the bleak hills in the West of England—Dartmoor, to wit—or on marshy, undrained land which is often enveloped in a dense fog, or elsewhere in a damp atmosphere. If tried in such situations, I opine, our horses would degenerate, and in a few years be no better than they were centuries ago. In Iceland and the cold countries, we are told, the horse is diminutive though strong. In Flanders, on the other hand, and in the South of France and other warm and congenial countries where the climate is dry, it is much larger. The climatic effect on dogs is, according to Goldsmith, still more extraordinary, almost surpassing belief, for he says, 'This animal,' speaking of the hound, 'transported into Spain and Barbary, where the hair of all quadrupeds becomes soft and long, will be there converted into land-spaniels and water-spaniels, and these of different sizes.' 'The mastiff,' he continues, 'when transported into Denmark, becomes the little Danish dog; and this little Danish dog, sent into the tropical and warm climates, becomes the animal called the Turkish dog, without hair. All these races, with their varieties, are produced by the influence of climate, joined to the different food, education, and shelter which they have received amongst mankind.'

We have the authority of this great student of natural history for the fact that these extraordinary changes are thus produced. Coming nearer to our subject, in writing of cattle and horses he expresses an opinion which confirms my own experience, and says that, 'Like our race-horses, we have indisputably the best breed of cattle and sheep in the world, which are eagerly sought for by foreigners, and transported to the Continent at enormous prices. We have undoubtedly the best breed of horned cattle of any in Europe, so it was not without the same assiduity that we came to excel in these as in our horses. The breed of cows has been entirely improved by a foreign mixture properly adapted to supply the imperfections of our own. Such as are purely British are far inferior in size to those on many parts of the Continent. But those which we have improved by far excel all others; our Lincolnshire kind derive their size from the Holstein breed, and the large hornless cattle that are bred in some parts of England came originally from Poland. We were once famous for a wild breed of these animals, but these have long since been worn out.'

If Goldsmith was able to write in such glowing

terms of the breed of cattle existing in his day, what would he not have said, I am tempted to ask, in praise of such animals as are now produced by the aid of skilful treatment and intelligent crossing? In respect to the wild breed of cattle for which we were once famous, I need scarcely say that the species has still some representatives in Chillingham Park. 'Druid,' writing in 1870, gives a description of them. It is really so interesting that I am sure a little space given to it will not be grudged.

'The herd,' he says, 'is generally kept up to 11 bulls, 17 steers, and 32 females, or three score in all. They are made steers of even up to four years old, and it is found even at that stage to improve the beef. It was the practice to do so when they were dropped; but it was a very dangerous one, and spoilt the bull selection as well. They are tempted into a yard with hay, and there snared, and tied by the neck and horn during the process, and returned next day without any cautery. The steers always grow larger horns and weigh from 40 stone to 50 stone of 14 lb. If it is fair weather they go up the hill, and if stormy they remain below. They eat very much at night, and mostly in company, and often scour a good deal in warm weather. The bulls are more of a tawny shade than the cows, as they fling the dirt very much over their shoulders when they kneel to challenge. Both sexes have black nostrils, horns tipped with black, and a little red within the ears; and in their general look they partake of the Charolais and Highlander combined. Their sense of smell is exceedingly acute, and a cow has been seen to run a man's foot like a sleuth-hound, when he had run for his life to a tree. While Sir Edwin Landseer was taking sketches for his celebrated pictures, the herd went into action, and he was glad to fly to the forest as they passed by.'

In many cases the improvement noted by Goldsmith now more than a century since (1774) has progressed and is progressing. Fifty years ago the rate of advance in this respect was a subject of congratulation to our forefathers. I may conclude that, in Goldsmith's time, the Shorthorn, Hereford, and other large breeds, as well as the beautiful Devon and Sussex, with many others of a smaller description, were not in fashion, or were unknown, or most likely non-existent, or they would assuredly have been mentioned by him. As a naturalist he would certainly, I think, have noticed and described their magnificent size, amazing beauty and perfect shape, as well as their extraordinary meat-producing qualities; and he probably would have informed us as to the breed of the stock from which they were originally derived, so that we might have perpetuated one so desirable.

The fact of continuous improvement since his

time is fully borne out by the independent testimony of later writers. 'Druid,' in 'Saddle and Sirloin,' speaking of sheep, says : ' Lord Polwarth's rams, as well as those of a few other flockmasters, were sold by auction at home for many years. In 1846 the Kelso public sales were established on the second Thursday in September, and 350 rams were entered, but £13 was the highest price. Lord Polwarth's were first brought to Kelso in 1852. In 1820 his lordship's home-average had only been £3 15s. for 35 ; whereas in 1865 it was £37 18s. 10½d. at Kelso for the same number. His lordship's top sheep went for £95 that year, and for £106 in 1867.' Again, he says : 'Two crosses of Cheviot have increased the Welsh sheep from 40 lb. dead weight (*i.e.*, the carcase without the head or legs from the knee, when the farmers sell by so much per lb.) to 70 lb.' (I may observe that fed on hay and turnips they have reached 90 lb.), 'and have also doubled the wool, on which the second cross seems to have good effect.' Referring to results in different parts of the Principality, he goes on to say : 'What has been said about upper Radnorshire applies as much to the higher parts of Montgomeryshire and Cardigan, but with this exception, that the Cardigan wethers seldom go to a fair. Many of them are bought for parks, and improve amazingly on the 5 lb. to 6 lb. per quarter which they would weigh on their arrival.'

So much for sheep. As for the improvement in cattle, what would our forefathers have said to an ox whose living weight was 216 stone of 14 lb. to the stone ? If we want to see what the intelligent application of knowledge, in suiting locality and treatment to the nature of an animal, has done, let us compare this weight with that of the wild Chillingham breed, which we learn is usually between 40 and 50 stone ! They had their marvels in those days, too. *Comet* is spoken of as 'the most symmetrical bull they had ever seen. He was not very large, but with that infallible sign of constitution—a good wide frontlet, a fine placid eye, a well-filled twist, and an undeniable back. His price caused breeders everywhere to prick up their ears. They had already heard of Fowler refusing 1,000 guineas for a longhorn bull and three cows, as well as for a cow and her produce of eight seasons ; but never of one bull achieving that sum.' But what are these sums, extraordinary as they are, to the fabulous prices that we see given now for all sorts of horned cattle ? And so with other animals, horses in particular, not only has the value increased, but the demand for them from all parts of the world.

Tempting as it is to follow up this interesting subject, I must not dwell any longer upon it. It is more material to our point to observe that extended knowledge and indomitable perseverance could not have achieved the improvement, without the advantages of a climate adapted to develop the good, and eradicate the weak points in the individual animals. It is, therefore, very needful on the part of the intending breeder to make sure that the situation selected has not the disadvantage of being either too bleak on the one hand, or, on the other, a marshy, undrained soil, and consequently damp atmosphere. There are plenty of eligible sites in all counties; and, therefore, it should not be difficult to secure freedom from these conditions; for the process of breeding horses subject to them must, I am sure, turn out disappointing in its results. There is another point. The reader of my former works needs no telling how essential I consider it to be that any horse, bred for sale for any purpose, should have both good feet and legs. The ground on which young stock is to be reared should, therefore, not be too hard. This, however, is scarcely likely to be the case, for, of necessity, a farm used for the purpose would have plenty of good pasturage, as it would not pay to rear ordinary stock in the expensive way in which thoroughbred stock is supplied with artificial food. This is a question that I shall deal with when I come to describe the stud-farm. And on this and other points I may refer to that description. One of these matters, of no little importance, is the fencing off or uprooting of the trees which I shall enumerate, the leaves of which are harmful if eaten.

For the rest, the selection of a suitable site may, I think, with these hints, be left to the individual; for, as I have observed, it is not so much a question of finding a locality—most farms being suitable for the purpose—as of taking the precaution of seeing that this or the other site is not too bleak on the one hand, or too humid on the other. But there is one point that I should not pass by; and that is the very simple question of room, for there is little use attempting to breed horses unless you can give them sufficient space in which to exercise themselves thoroughly.

This question of exercise, I think, is often not sufficiently considered. For one thing, you can hardly overdo a colt with food if he has only liberty and space for exercise. When I come, in its proper place, to the treatment of young horses, I shall have something to say on one grand mistake made in their management, and that is, insufficiency of exercise through being kept in small paddocks. But I might here, whilst I am on the subject, relate an interesting experience of my own in the case of some greyhounds, bearing on the matter and illustrating it.

Good or bad action, it must be remembered, is a very material point in the value for sale or actual use of any horse. Good action, as I shall presently show, is undoubtedly inherited. But it is also developed to a considerable extent by opportunities of abundant exercise in youth. I lately saw an extraordinary case of this sort in two well-bred five-months-old greyhound puppies, out of Mr. Randoll's bitch *Rona*. They were generally fed and kept in a loose-box, and allowed to take exercise on the lawn and grass fields close by. In this way, whilst playing, they would chase each other with so much vigour and determination that they soon lay down fairly exhausted, with their tongues out of their mouths, gasping for breath ; in fact, they would tire themselves as much as if they had actually had a short course after a rabbit or leveret. This sort of exercise they took most days when the weather was fine. They grew up to a good size, and had the best of attention, and are now promising saplings. Mr. Randoll had also another bitch puppy of the same litter, which, for want of room, was treated differently. She was, as may be readily supposed, the finest of the three, or such a good judge of greyhounds would not have selected her for himself as he did. She was kept in a small stable at Salisbury, well fed, and in every other respect well cared for, yet she did not thrive. Mr. Randoll thought it must be from confinement, since he could not let her out for exercise as often as he wished, for fear of losing her in a crowded town. One day, seeing mine doing so much better than his, he asked me to take the other sister, which I did, and they were all treated alike at Coombe. The new-comer was hearty, and ate as well as the others; but her action was like that of a high-stepping park-hack, and she galloped as high as a rocking-horse, and was 'as slow as a man in jack-boots,' to use a colloquial term. In fact, when running 50 or 100 yards, a distance they often did as fast as they could back to me, when they found I had turned round unobserved by them (which I often did purposely, for the sake of seeing their relative speed), she was invariably left to perform half the distance by herself. It was not for some months after that any improvement was made in her action; but with attention and exercise came an increase of speed, and though not nearly so good as the other two, yet her action greatly improved. A few months later she took the distemper, and gave it to the others, from the effects of which she died; but the others recovered. Does not this suggest, or, properly speaking, teach us the absolute necessity for all young animals to have their freedom and the power of exercising themselves at an early age, as often and to what extent they choose, in large spaces that will enable them to extend themselves without restraint, for the sake of their action as well as for the benefit of their health, and, I may add, for the saving of the vet.'s fee and the bill for nauseous drugs?

CHAPTER IX.

SLIPPING AND STERILITY.

Wide extent of loss from this cause—How ewes are treated—My experience shows that it is a disease—Early exhibited at the stud—Twins also a loss—Inherited and accidental causes.

Nature of malady examined—Examples through three generations—Dead foals lead to barrenness; examples—Afflicted mares die early—Slipping leads to barrenness; examples—Extent of malady shown in 'Stud-Book'—Examples from one hundred representative cases—Must end in loss.

Accidental causes—Care needed during pregnancy—Removal of refuse in paddock and yard—Fright—Bad smells—Must separate barren mares—Extraordinary instance at Alvediston from eating fallen leaves.

Result of examination—Slipping and sterility go together—Exemption of Irish mares suggests the question of climate and treatment.

SLIPPING and sterility are fruitful and never-failing sources of disappointment and loss to all breeders of stock. They are not peculiar to the equine race, for we find them existing in flocks and herds in all parts of the country, so that they should be jealously watched by all owners of sheep and cattle, of whatever breed or description. If a ewe casts her lamb she is immediately marked; and at the end of the season, when the surplus stock is sold, she has to go in company with those which have proved barren, or have had bad udders, or any other defect likely to prevent them from breeding and rearing their stock in good condition.

The same strict rules are followed in the case of cattle. Cows which do not give satisfaction in this respect are, like ewes, got rid of without consideration of price, by breeders who wish to eradicate the disease of sterility from their herds. Here we have a good and sufficient reason for the application of a similar rule to brood-mares. It is a subject which has always occupied my attention, and I think that an account of my own experience in the matter should be useful to breeders of horses of all kinds. For, although my lesson has been learned chiefly in raising thoroughbred stock, yet I think its teaching applies to mares of all kinds, and especially in the fact that I have been able to trace the causes of the disease to certain sources, and, therefore, am enabled to suggest remedies that, in my own experience, have proved successful.

That slipping is a disease cannot be doubted. Nor can it be doubted that, to a great extent, it is hereditary. Like other animals, it may be said of brood-mares that many of them have a predilection for doing certain things, and this predilection is generally shown at the commencement of their life at the stud. For instance, if, in her first year or two, a mare should breed a winner, though only of a little race, it is a good omen. Better things may reasonably be expected of her. So with mares predisposed to sterility, the defect will be discovered in the first year or two, as a rule; and in the same way, mares disposed to slip their foals will exhibit the tendency at the same early period. There is nothing for it, in my opinion, therefore, but to discard from your stud any animal that commences her career by slipping or having dead foals. The same objection and process, I should add, applies to those that have twins, as these, for racing purposes, or indeed for other purposes, are generally useless. These facts, for facts they are, though they may have been unobserved by some and disregarded by others, should teach us the necessity of seeking the origin of the disease, with a view to its remedy; for if we know the cause of any malady, even the most obstinate may be 'considered half cured.'

For my own part, I am convinced that slipping and sterility are grave defects which, like most other diseases, are generally inherited; although, as in other defects, there are also accidental causes, for it is an ascertained fact, and one accepted by the most attentive and intelligent breeders and veterinary surgeons, that there are many exciting causes which produce abortion. It will therefore be a convenient course to consider first the innate origin of the complaint; and secondly, the causes which accidentally produce it, so as to guard against the occurrence so far as it may be possible to do so.

As to the hereditary nature of the malady, I have found that, when mares slipped a greater number of foals than the generality usually do, their offspring had a greater tendency to do the same thing than others have which have been bred from mares not so afflicted. Of the truth of this, and of the difficulty of overcoming it—and, I may add, the loss resulting from breeding from a stock predisposed to it—the following illustrations will, I think, afford ample proof.

Octaviana, the dam of *Crucifix*, slipped. *Crucifix* did the same. *Chalice*, out of *Crucifix*, in addition to slipping a twin and a single foal, was barren five times and had two dead foals. *Faith*, by *Pelion* out of *Chalice*, had two dead foals, and was barren the next year and for four years subsequently. *Reigning Beauty*, by *King Tom* out of *Chalice*, had two foals, was barren two years, and was given away. *Verona* slipped, and so did her daughter *Villafranca*. The same was the case with *Cerise* and her daughter *Pinkie*. All these mares, in short, were not only barren, but slipped also.

If these instances do not prove that the disease is hereditary, I cannot think what further evidence could do so. Here are three generations clearly traced, in which the mares were barren or had dead foals. And the tendency is shown, too, in the collateral branches, in the two near relatives which slipped in the first year.

The same result follows with mares that breed dead foals, or which give birth to foals that live only a short time, as the following instances show :

Torment was barren the first year. Her first foal died as a yearling ; she subsequently had two dead foals, and was barren in all nine years. *F sharp*, her foal, a twin, was barren the first year, had a foal the second year, and was sent to Brunswick. *Tormentor*, another of her daughters, had a dead foal the first year, was barren the next year, and for four years afterwards. *Tisiphone*, out of *Torment*, had four foals, all of which died young, and, like her mother, was barren six years.

Monstrosity, dam of *The Ugly Buck*, slipped twice, and was barren four years, and died a young mare in 1857.

Marigold slipped three times and was barren the like number of years. Her daughter, *Blue Rose*, was barren the first year, and died very soon afterwards.

Legerdemain slipped a filly foal the day after winning the Cesarewitch, was trained for two years after, and again slipped twins in 1856, 1860 and 1862, being barren for three years.

Miserrima slipped three times, had dead twins

twice, one dead foal, and was barren five years. *Mœstissima*, her daughter, had a dead foal, slipped, was barren, and died young.

I find, indeed, that most mares have slipped or had dead foals before having twins. It was the case with *Device, Caricature,* and *Advice*.

Mandragora is another and a striking example of the hereditary character of the disease. She had four dead foals, and produced nothing for seven years. Her daughter was still worse, for she only bred two living foals, was barren twice, and died a young mare.

With such examples before me, I should be very chary indeed of breeding from a mare that had either slipped foals or had dead ones, or had been barren during the first two years at the stud. Furthermore, it may be said, and I think with truth, that mares which bring forth their young immaturely seldom live so long as those which go the full time and produce healthy offspring. Besides, many mares die young after slipping their foals. Nor is this experience confined to horses; the same may be said of animals of every description. And beyond this danger to life, I have noticed that mares which have slipped are barren twice as often as those which have not been so afflicted, and in most cases have more dead foals. I may mention, as an example, that *Handicraft* slipped the first year, and was barren the next four years. Amongst other mares I could name, which were either barren or slipped in the first two years at the stud, and subsequently were either barren, had dead foals, or slipped oftener than most mares, were *Happy Wife* (barren the second year, five times afterwards, and slipped), *Vicar's Daughter, Botany Bay, Silvia, Rapidan, Gift, Marie Stuart, Minna Troil, Isilia, Hampshire Lass*, and two I have already mentioned, *Verona* and *Marigold*.

The importance of the subject must be my excuse for dwelling upon it. It demands, indeed, more than an isolated case to prove the fact to my own and the reader's satisfaction. To show to what extent the injury from slipping and sterility really affects the profits on breeding, I may mention that Messrs. Weatherby give the following particulars of the results for 1879 : out of 2,860 thoroughbred mares put to the horse, 906, or nearly one-third, were barren or slipped. In 1882, 2,874 mares put to thoroughbred stallions produced 1,916 foals ; 818 mares were barren, and 140 slipped, or one-third had no foals.

To be more fully confirmed in my view, I have, at considerable trouble, gone through the fourteenth volume of the 'Stud-Book' to ascertain the names of a number of mares that have either slipped, or had foals dead, or that died shortly after, or twins. And the result simply proves that my theory—if what has been learned through experience can be called theory—is absolutely correct; for, except in two or three cases, which go to prove the rule, these mares were barren also, either before or after such misfortunes, and more often so than is the case with the generality of brood mares. Of these, I have taken one hundred cases indiscriminately, in which they are described not only as having been barren, but barren much oftener in a given number of years than most other mares. Of these hundred cases in which the mares have slipped, I will give a few instances:

Tape, I find, slipped three times, and was barren five years.

Algerie slipped foals in 1878, had twins in 1879, and died the next year.

Red Leaf slipped once, then was barren for three years, having had two living foals before.

Raffle slipped three times, and was barren a like number of years.

Silvia, a mare I once had, brought forth her first foal dead; had afterwards two more, likewise dead, and was barren for eight years.

Verona slipped twins once, was barren seven years, had two dead foals, and died after foaling.

Abigail was barren three years before and three years after she bred *Ruby*, her first foal, and was again barren four years subsequently.

Worthless slipped twins in 1870, cast her foal in 1877, and died after foaling twins in 1880. She was not barren, however; although slipping three times, twice with twins, is worse than doing so once and being barren twice. Even so, she can scarcely be taken as an exception to the general rule, for had she lived to the age of many of the others, she, like the majority of them, might also have been barren.

Omicron, to add one other example, slipped three times, and was barren three years.

I could add innumerable other instances, but these, I think, will be sufficient to prove my proposition, that when mares, in their first or second year, give birth to premature or dead foals, or are barren, they should be got rid of at the earliest opportunity. It can hardly pay to keep even good mares at the stud if they do not breed oftener than once in six or seven years, and then most likely have little or weakly foals.

As to the causes which may produce abortion by accident—for that such exciting causes exist, is, as I have said, the opinion of those who are best entitled to speak on the subject—they are various, such, for instance, as the kick of a vicious companion, or through eating cold grass before the hoar-frost has disappeared. Treatment, therefore, has something to do with it. And, amongst other things, great care should be taken not to expose a pregnant mare to any unpleasant smell, such as that of carrion, or even blood, as it often produces abortion, and the mare once so affected is very liable to be so again ; and whether sympathy or some unknown circumstances be the exciting cause, she may induce many of the other mares in the same paddock with her to cast their foals also. Of the truth of this assertion I think there can be no doubt, as it has been proved over and over again, and deservedly demands our most serious attention.

Sudden fright, such, for instance, as the discharge of fire-arms near the stable or in the paddock would cause ; bad drains in or near the stable ; pestilential vapours rising from putrid matter, or from the decaying vegetable substances always to be found in dung-yards during the autumnal months, are actively potent causes. The dung-pits, indeed, should never be situated too close to the stables, or the dung should be frequently removed on to the land, so that the yard may be kept free from obnoxious smells. I believe many mares may and do cast their foals occasionally from other causes ; and one is, when barren mares are allowed to occupy the same enclosure as those that are in foal—the restless disposition and galloping propensities of the former often disturb the quiet and repose of the latter, making them exercise themselves beyond their enfeebled strength.

Any barren mares should be removed, therefore, from those that are in foal and kept in a paddock by themselves. This should always be done.

Again, mares are sometimes kept in enclosures where noxious herbs and trees abound, the leaves of which, when eaten, will produce abortion. Of this fact I can give an astounding case, which may be fittingly related here. In a paddock of about four acres, entirely surrounded by trees, at Alvediston, I kept a few (some six or seven) broodmares during the autumn months in the day-time. The soil was rich and the herbage rank, and, strange to say, nearly if not the whole of them slipped or were barren. There were no yew-trees in or near the place, but a splendid avenue of limes, which, like other deciduous trees, such as the elm and ash, were scattering their leaves in all directions, covering the ground. To the eating of these leaves I always attributed this otherwise unaccountable fatality. In after years I took care that the mares in foal were never allowed to enter that paddock in the autumnal months, nor any other where the leaves could be scattered by the wind in great numbers, and I never had the recurrence of such a calamity. In fact, slipping amongst my mares, forty or fifty in number, was a rarity; and I am, therefore, the more assured in my mind, as I was at the time, that the cause was to be found in their eating the fallen leaves, principally from the lime-trees, and perhaps a few of the lower branches of the trees which they could reach.

I hope this somewhat lengthy examination of one subject has not been without a useful result. I have shown, at all events, how extended and disastrous in its effects is this want of fertility. And I think I have also shown that, in two ways, the chances of a mare slipping her foal may be minimized, or at least greatly reduced; for its absolute prevention is not to be expected. Slipping and sterility, it is seen, generally go together, and though my examples have been drawn from thoroughbred stock, it is fair to presume that the facts elicited apply with equal cogency to broodmares of every description. Indeed, half-bred and other mares have not the same care taken of them as is bestowed on the thoroughbred, and, therefore, with them the mishap may oftener occur from preventible, because unnoticed, causes. This is a matter for serious consideration. The annual loss from this infirmity to breeders and the country is incalculable, and it therefore behoves all interested persons to look well to the soundness of the animals they breed from. For, most undoubtedly, unsound and diseased mares breed more dead or sickly foals, or more often slip or are barren, than such as are constitutionally sound. And beyond the careful choice of the brood-mare in this respect, there are the accidental causes of abortion, to some of which

I have referred. Here, care would do away with the evil, prevent slipping, and save the life of many a mare.

I should say, in conclusion, that in my examination of the results shown in the 'Stud-Book,' I found, in going through the list of the Irish mares, that they are less prone to this disease than ours, although barrenness followed much in the same ratio as with our own. From this it would appear that the climate or the land is better adapted for the purpose of breeding than it is here, or that they have a better system of management than is pursued in this country. The fact may be worthy of consideration by intending breeders, for anyone who found out the cause of mares slipping or breeding dead foals, and produced a remedy, would be hailed as a benefactor to his country, and deservedly so too.

CHAPTER X.

TEMPER, SIGHT, ROARING, SPLINT.

Temper of horses important—Traced through generations—*Young Trumpeter* and his offspring; *Little Tom* and *Bugler, Sigmaphone*—Narrow escape of *Crucifix*; practical loss of her second race; always started badly; evil recurrent in *Surplice*—Examples of complete failures; *Entre Nous* and *Vittoria*—Defects due to temper—Danger in the paddock—Effect of improper punishment—Influence of breaking on temper; an opinion on the Sledmere process and my own opinion—Mr. Farquharson's ill-success.

Faint-heart a source of great loss—Examples in *Allie Slade* and others.

Defective sight, transmitted and accidental—*Drogheda* and *Defence*—Cause of accident and expense; instance with Mr. Sadler—Inherited nature of disease proved by decrease in cataract.

Roaring a transmitted disease; often dormant; instance in *Ormonde*—Supposed remedies—Influence of climate; *Belladrum* in South Africa—Lessened by better treatment.

Splints, if serious, may be inherited—Need to discern between accidental and inherited maladies.

WE may now conveniently consider the consequences of some other defects which are chiefly inherited, although occasionally the result of accident or improper treatment—defects which are common to horses of every description, and are not indicated by their shape.

First amongst these I put TEMPER. No one who thinks the matter over would, I am sure, for a moment contemplate breeding from a bad-tempered animal, and so perpetuate a savage race of unreliable horses. This is not merely a question with the race-horse, so as to secure an animal that may be depended upon to do its best, running one day pretty much as it would do on another; but it applies with equal force to carriage-horses, hunters, and hacks. That this is so needs no argument. Many a human life is lost through the uncontrollable temper of a vicious horse, whilst accidents to horse and man alike daily arise from the same prolific cause.

Temper, like colour and action, may be traced back to its origin through many generations. This is a point that very frequently is not sufficiently considered. If we, from our own knowledge, are satisfied with the temper of a stallion or mare that we have seen run a few times, we dive no farther into particulars, though either on the paternal or maternal side, but a few generations back, may be traced a temper the most vicious, by those who are more than superficial observers. Here we look at the parents and lose sight of their ancestors, which the offspring may favour more than their immediate progenitors, and a vicious brute is the disappointing result.

I may take, as a first example, *Young Trumpeter*,

who was never trained, owing, it is said (and probably so), to an accident when young, and was early put to the stud. He got his strain of temper through *Surplice* from *Crucifix*. He had but few mares put to him, and those mostly bad ones; and yet nearly the whole of his stock could run. Two of his sons showed a marked difference. *Little Tom* was backed for the Cesarewitch and Cambridgeshire Stakes, in which, carrying only 5 st. 10 lb., he ran badly; although between these two races he won a Welter Plate, carrying 8 st. 5 lb.—but he never won a race again. *Bugler*, on the other hand, of the same age, was nearly the best two-year-old of his year, an offer of 2,000 guineas being made and refused for him. Yet in private *Little Tom* was always the better of the two. Many other horses were got by him, and nearly every one was more or less bad-tempered, amongst them being *Orchestra*, who won a race at Salisbury, but never won another. One more of his produce, *Sigmaphone*, was as fine a horse as his father, and a good two-year-old, having at that age, at Stockbridge, been tried with *St. Blaise*, and is said to have beaten him; and the year following he (*St. Blaise*) won the Derby. But *Sigmaphone*, after winning one or two races as a two-year-old, never won after. His temper was so bad that he could only be mounted with much difficulty, if not actual danger. Luckily, for the sake of the turf, *Young Trumpeter* did not live long.

Crucifix herself was a bad-tempered mare. In fact, her temper on one occasion nearly cost her her life, and imperilled that of her rider. For when sweating she became unmanageable, bolted, and fell over a hedge and ditch into a plantation. Her temper really lost her second race; although, by a fluke I may say, it was ultimately adjudged to her. This was in the Chesterfield Stakes, in which she was fairly beaten by an animal she could give two stone to. The start, however, on the evidence of the starter, was overruled, and on the second time of asking she won in a canter. In most of her races she never started until the other competitors had left the post. For the Criterion, the start was delayed for a very long time by her fractiousness, which lost her so much ground that she only got up in the last few strides and made a dead heat of a race she ought to have won. The display she made in running for the Oaks was little or no better. Her temper, I may say, her son *Surplice* thoroughly inherited, as was clearly shown in the difference of his running at Goodwood and his running just previously and subsequently at Epsom and Doncaster respectively. He was the sire of *Eugenie*, the dam of *Young Trumpeter;* and we may conclude that if the latter had been raced, he would have run as deceitfully as his progenitors, if we are to judge by the stock he left behind him.

Entre Nous is another example to the point.

Had I known as much then on the subject as I do now, or had I thought better of it before I bought her, I might have saved myself much useless expenditure. I bought her as a yearling. Even then she had the most unmanageable temper, and at that tender age scarcely anything could be done with her. As a two-year-old she seemed better, and was tried a good mare; but in running her temper was as bad as ever. After this I sold her, and she ran no better when backed by her new owner. All this, if I had thought of it before, might have been expected, for she was got by *Landmark,* most of whose stock were as roguish as himself.

Vittoria, by *Arthur Wellesley* out of *Dewdrop,* was, if possible, a still more glaring case. She was tried a good mare, but ran several times in bad company as a two-year-old without success. The following year she ran third to *Controversy* and *Brigg Boy* for the Lincolnshire Handicap. In this race, which was probably the only race in which she tried, she showed at least two stone improvement on any form she had previously exhibited. Her real temper was, however, displayed in the same decided manner the week following at Epsom, where she was nearly last. I afterwards sold her to Mr. Pattison, for whom she ran many races, but was never honoured with a bracket; and she, like *Entre Nous,* was put to the stud, and their stock have not yet, nor are they likely to, hurt anyone but their most sanguine owners. *Falmouth*, out of the same mare by *Glenlyon*, was sold privately by Mr. R. Wright to the late Mr. Joseph Dawson for £100; he ran and won when a two-year-old, became a favourite for the Derby, and was, I believe, sold for £6,000; but, like his half-sister, he displayed the worst of tempers, and never won a race after.

I do not for a moment say or believe that every unsuccessful attempt at winning can fairly be ascribed to the account of temper, for I know well enough that such is not the case. But I am thoroughly convinced that many defects can only be due to horses having an uncontrollable will of their own, and this is one reason amongst others to warn people against breeding from ill-tempered animals.

There are other objections, as I have said. For one thing, bad-tempered mares, when loose in the paddock with other mares and foals, are dangerous, as they frequently hurt themselves or their companions through their vicious propensities, often savagely chasing the others. This is more frequent in small enclosures than in large open paddocks; but in either a mare or foal is sometimes killed in this way, and more often rendered useless through injury.

Temper, being inherent in the horse, as I have shown, may be easily developed by improper treatment. That ill-usage will produce it, such as an unprovoked beating or undue punishment for some trivial fault, I need scarcely say. This, however, generally occurs after the animal has left the breeder's care, and is not a matter of such moment to him, because temper thus artificially created will not so certainly affect the offspring. There is, nevertheless, one aspect in which it has an interest of its own, and that is the question as to how far breaking at an early age has an influence on the temper.

This is a subject that more properly, perhaps, comes under the head of 'treatment,' and therefore should be dealt with when we come later to consider the different methods pursued. However, as it has so much to do with temper, I think I may be excused if I introduce it here. On this point I should say that I have consulted a gentleman of great experience and ability as to whether horses should be broken and raced before being bred from, as many of the late Sir Tatton Sykes' and a few other breeders' formerly were. He writes in reply :

'No mare should be bred from without being properly broken and ridden, for two reasons. The frame of an unbroken mare never develops, and if the temper is not made subservient by the process of breaking, the produce is so much the harder to deal with. Both these statements the late Sir Tatton Sykes proved for the benefit of subsequent breeders, inasmuch, for generations the mares had never been broken, though kept well enough, as I had ample opportunity of seeing, and consequently they grew every year smaller, and, considering the number of horses he bred, no man ever bred so many bad ones, and sometimes their produce were perfectly intractable when put into work.'

I think the writer must be speaking of many years ago, for I had a great many of Sir Tatton's yearlings at one time or another, and I am bound to say I did not find them more difficult to break than others at that age. But I think that of late years the baronet had many of his mares broken and ridden, though they may not have run in races. This would account for their produce not having worse tempers than most other yearlings. Nevertheless, I can thoroughly confirm this gentleman's opinion, by a quotation of my own from 'The Racehorse in Training,' where I say :

'The late Mr. J. J. Farquharson, of hunting celebrity, used to keep his horses till four and even five years old running loose in large paddocks before sending them to be broken. But I never knew him possessed of a good one so treated. A winner he may have had, which is all that may fairly be said in favour of the produce of his stud. A more savage lot of horses in the stable I never saw, or on the turf a greater set of rogues. This experience should, I think, deter anyone from following a plan that, in my opinion, has so signally failed, and which may be said to be the only one tried to such an extent without a redeeming point.'

In conclusion, I may say on this point that all horses and mares should be ridden and made quiet before putting them to the stud, if only for the improvement that it may reasonably be expected will be made in their temper by judicious and early restraints. Good-tempered mares, and good-tempered mares only, should undoubtedly form the stud, for in many, if not most cases, the offspring inherits the failings, just as it inherits the other peculiarities, of the dam. A vicious mare may breed by accident or some freak of nature a good-tempered animal; but it would be an accident, and the old inherent nature, like an incurable disease, would not be got rid of, but would show itself when least expected, and consequently with all the more disastrous effect.

FAINT-HEART is another not outwardly visible defect, which should be guarded against in selecting brood-mares or sires, for it is clearly one that is inherited. Unfortunate animals afflicted with want of courage are most disappointing to their owners. At times they will beat anything that may be brought against them, whether in trials or races, and by these deceptive performances will induce their unlucky owners to half-ruin themselves by backing them for important events, in which they do not run as well, because they will scarcely put one leg before the other.

Stuff and Nonsense was such a one, many of her foals being currish. She was the dam of a mare by *Typhœus*, bred in 1874, which was always thought a good mare in private, but could beat nothing in public after her first race or two. The year following she bred *Allie Slade*, an own sister, who was affected pretty much in the same way; for although she managed to win a selling-race at Egham, as a two-year-old, and though very sound, she never won after, and the gentleman into whose hands she fell told me that she could not beat a hack in private, and retired from the turf for the paddock without ever running in public again. Now, this mare had no signs of a bad temper. She was quiet to ride, and never offered to bolt or run away with a lad at exercise, and in the stable was as docile and tractable as a lamb. But in her races she had no heart, and lost them simply because she would not struggle for victory at the critical moment.

DEFECTIVE SIGHT is another evil more often transmitted than acquired. No one, I imagine, would dream of breeding from a mare or horse that had not good eyesight, unless it were a well-known fact that the loss of sight was the result of accident or of design, as in the case of *Drogheda*, whose sight was intentionally destroyed by throwing vitriol in his eyes—a cruel process resorted to on account of his savagery, as nothing could be done with him so long as he could see.

But that blindness is transmitted to the offspring has been only too clearly proved, and that the risk is too great to be run. Many of the *Defence* blood were so afflicted. He was himself blind, and so was his son *Safeguard*. His daughter *Defenceless* was also blind. She was the dam of *Caractacus*, winner of the Derby, and had he been similarly afflicted, and of course the probability was great that he might have been, what a loss it would have been to his owner, Mr. Snewing! I do not rely on this individual case of blindness in a family; there are others, and not a few, which show that blindness is transmitted, and, therefore, I should on no account like to put my mares to a blind horse, or breed from a mare with defective eyesight, for it, like all other hereditary diseases, may be traced back for many generations.

An instance of the inconvenience and danger of keeping such animals occurred to old Mr. Sadler, who had one of his best mares, but unfortunately blind, feeding in a paddock opposite Pyrrhus Cottage, Stockbridge. She fell over the embankment, which is very steep at this point, into the road below, and broke her neck, being at the time heavy in foal. I believe that she was rubbing herself or leaning against the rail that was put up for her protection, when it suddenly gave way. There is another objection to breeding from animals so afflicted, and that is the considerable extra trouble and expense they cause, from having to be kept by themselves in separate and secure paddocks and boxes specially fitted up for their safety.

As to the possibility of lessening the malady, by eschewing for stud purposes all horses showing a tendency to it, I may add that we have it on the highest veterinary authority that 'for one case of cataract we see at the present day, we used thirty years ago to meet with twenty, or even with more than that.' 'Why? Because it was the disease which, perchance, first roused the suspicions of the breeder as to its hereditary character, and warned him that to breed from blind animals was a mistake.'

ROARING, notwithstanding all that may be said to the contrary, is undoubtedly, in many instances, a disease that may be, and often is, transferred from the parents on either side to their offspring; and nothing would induce me to breed from an animal afflicted with such an infirmity. We have seen this in numerous cases when the mare, a roarer, has bred a colt inheriting the disease of his dam, who at the stud in turn begets a race of musicians, which, like the drummer at a country fair, are more noted for noise than harmony.

A tendency to the disease is not always discoverable by merely ascertaining what has been the health of one generation. Like the gout, it is often known to lie dormant a generation or two, only to reappear in another with increased violence. Even when the infirmity arises from illness, as in horses running for years and then becoming roarers, it is just as well to avoid them for stud purposes. I admit that in such cases, in which roaring has not been known to exist in the family for a number of generations, to breed from such animals, if exceptionally good, may be excusable. But no matter how comparatively accidental may have been the exciting cause of the disease, if the disease has been in the progenitors, the animal should not be thought of for stud purposes. Of the extreme disappointment which comes to owners through the existence of this malady, dormant for a time though it be, we have a sufficiently striking example at the moment in the case of *Ormonde*, who, the best horse of his year, and probably of the decade, has just recently fallen under the ban of this complaint, much to the disappointment, it must be, of his noble owner, the Duke of Westminster. Here the dormant nature of the disease is only too clearly shown; for *Pocahontas* was a roarer, and her first foal *Cambaules* also; yet many of her sons and daughters escaped the infirmity, only for it to reappear in *Ormonde*, a relation further removed Moreover, some of *Ormonde's* family on the dam's side were roarers.

Science has lately brought to light much that was previously unknown as to the cause of this disease. Unfortunately the advance does not extend to the discovery of a remedy for it: and for all practical purposes we might as well be without the knowledge; for knowledge, if it cannot be applied, is to all intents and purposes useless. In short, so far as remedy is concerned, the true friend of the horse and his breeder can only advocate that every roarer, and indeed every descendant of a roarer, should either be castrated, or be sent to the Cape of Good Hope, where roaring is unknown, and no harm therefore could be inflicted on the native horses.

But we have learned something more practicable in the prevention of the disease by rule-of-thumb experience. For I certainly think there are not so many roarers now as formerly, and that this may be accounted for by the better treatment horses receive now than they did in years past, when, after racing all the season, they were turned into a loose-box without exercise for months, till taken up again and put into training for the following year's races; and when hunters, having been turned out to grass in the summer for soiling, were put suddenly to strong work and high feeding. This I consider to be one prolific cause of the disease.

As for the effects of climate, the late Mr. Merry's

Belladrum, once a great favourite for the Derby, was a bad roarer, as is well known, and was sent to South Africa, where, in a few months' time, he became as sound in his wind as any horse that had never been so afflicted. I had this from a gentleman who lately returned from the colony, who knew the horse and had seen him run in this country before his expatriation; and he assured me of the fact. Quite independently, however, of the result shown in this case, I have long been of opinion that horses are more likely to escape or recover from the malignant effects of the disease in a dry locality than in one that is damp, bordering on rivers, lakes, or stagnant pools; and, above all, than in stables kept at injuriously high temperature, as I have elsewhere observed. Firing and blistering the horse's throat, as introduced by the late Mr. Robinson, of Tamworth, and many other remedies, have been applied, but only with partial, if with any decided success. No doubt a horse so treated may get better, and a cure be therefore erroneously ascribed to the effect of the application, whilst really, as in the case of *Devil among the Tailors* and *Brigantine*, it would have recovered if left, as the two latter were, to nature's powerful sway.

SPLINT is another complaint that may be inherited; 'although,' says an eminent veterinary surgeon, 'I fancy I hear the reader exclaim, "If you object to breed from horses affected with splints, you would have no horses to breed from." In reply, I would observe that only in cases where both on the inner and outer surfaces of the legs, or on all four legs, conditions exist which fairly indicate a tendency in the animal's system to bony deposits—should I consider them a decided objection. It is argued by some that it is a good plan to so mate animals as to counteract a defect. For example, to put a sound-winded horse to a roarer; or a horse with curb to one with very good hocks. But in my opinion, as a rule, this does not succeed, for it will be found that, although the progeny may not in early life have the disease, there is a tendency to it—a weakness in the part which an exciting cause will sooner or later develop.'

' In conclusion, I should say that I do not wish it to be understood that every animal affected with any one of these diseases is to be avoided as unfit to breed from. I am well aware that each and all of them may arise from accidental causes. A careless carter, by dropping the shaft upon his horse's coronet, may produce a sidebone; a blow at the back of the hock may create an enlargement resembling a curb; an injury to the eye may cause cataract; and I believe that senility, or a very bad attack of strangles, may occasion roaring or whistling. I would, nevertheless, recommend that care should be taken to determine whether such injuries had been inflicted, or whether the seeds of the disease had not previously existed, and only awaited the exciting cause to bring about its development.' That is the opinion of Professor Pritchard, with which I may say I entirely agree.

CHAPTER XI.

CONSANGUINITY, CROSSING, COLOUR.

Difference of opinion on in-and-in breeding—Purity of blood essential; value of crossing for the purpose—Tendency of domesticated animals run wild to breed back—Early neglect of the horse—Darwin and Youatt on effect of selection in breeding—Darwin on crossing and fertility—Effect on the human race—The 'daft' villager—Loss of gameness in fowls—Greyhounds improved by cross with bull-dog—Effect on the racehorse—*Melbourne's* produce—The thoroughbred cross in all our horses.

Transmission of colour through a century—Marked instance in *Crab* and his offspring—Descent of *Chanticleer* stock—*Grey Friar*—*Birdcatcher's* grey flanks—Colour not material, but serves as proof of transmitted qualities from ancestors.

I PROPOSE now to consider the subject of consanguinity, concerning the effects of which there is some difference of opinion. There are breeders who have a great liking for close affinity in their stock; others, again, are as strongly opposed to it, believing it to be the origin of no little evil.

Amongst the fundamental principles in the art of breeding, there are some that cannot be transgressed with impunity, to which I have already referred. The necessity of good-temper is one. Another is freedom from constitutional or hereditary defects. Good action is another, to which I have briefly referred in insisting upon sufficient room for the proper exercise of the stock, young and old. I shall again recur to it when dealing with the selection of the mare and the treatment of the foal.

Purity of blood is another essential, and perhaps the most important of all; for if your object in breeding is to produce the best stock, you must start with animals of the purest blood. In this, as in other respects, we can find analogy in other animals. The best bred cattle, for example, always command the highest price in the market, when in other respects not defective; a fact well known to all breeders and dealers. A similar result is attained by following the same undeviating rule with the Southdown, Leicester, and other breeds of sheep. It is the same with pigs, and cannot be otherwise in breeding horses, whatever may be the sort it is proposed to raise.

That all our present breeds of horses sprang from common parents originally, I have no doubt. We have an apt illustration in ducks; all the different known sorts came from the wild duck, with which they will breed readily, and evince a desire to regain their original freedom. Pigeons, though we have several varieties, differing in size and colour, all came undoubtedly from the wild pigeon —the blue-rock, to which at the present day, in many respects, they bear a very striking resemblance. We have more marked examples of this nature in domesticated animals that escape from us and run wild; as we often to our cost know they do. These gradually but invariably revert to their aboriginal stock. For instance, cats are often known to run wild in woods and there breed a race of their own, which become in every particular like the wild breed. Again, tame rabbits, when turned loose, or those that escape from confinement, soon become as wild as those that have never been captured, the one breeding freely with the other. Once more, ferrets, when lost in the burrows of rabbits only for a few weeks, become as ferine as the pole-cat, with which they have been known to breed, and of which they are no doubt a species, whilst their offspring are totally irreclaimable, following the savage propensities of their progenitors.

These instances clearly prove the natural tendency of all animals to revert to their parent stock; and show that if left to themselves they will, by in-breeding, gradually lose all the qualities that may have been developed by care, climate, and selection for the purpose of mating. What attention in this respect will do, when only guided by uneducated instinct, we see in the case of many of the aborigines, who look to the breeding of their dogs whilst neglecting that of their horses. Indeed, we may conclude that, if it had been the practice to pay attention to the breeding of horses in early days, even amongst civilized people, we might have long since possessed good horses, perhaps as good as any we have now. In which case our present breed would be an improvement on those, for all science is progressive. The ancient peoples did not appear to exert themselves in improving the breed of their horses; as we have seen, higher qualities were gradually developed unconsciously, and from 'natural selection;' thus each successive generation was only by accident, to a great extent, an improvement on the one preceding it. It was not until assisted by man at a later period that the horse reached its present state of perfection.

I may here quote the late Professor Darwin on the subject:

'It is certain,' he says, 'that several of our eminent breeders have, even within a single lifetime, modified to a large extent the breed of cattle and sheep. In order fully to realize what they have done, it is almost necessary to read several of the many treatises devoted to this subject, and to inspect the animals. Breeders habitually speak of an animal's organization as something quite plastic, which they can model almost as they please.'

'Youatt, who was probably better acquainted with the works of agriculturists than almost any other individual, and who was himself a very good judge of an animal, speaks of the principle of selection as "that which enables the agriculturist not only to modify the character of his flock, but to change it altogether. It is the magician's wand, by means of which he may summon into life whatever form and mould he pleases."'

If we consider how the breed of the English horse has been improved in the last two hundred years, we can but arrive at the conclusion that the present condition has been reached mainly by judicious crossing, which has made the offspring superior in every quality to either the original dam or the original sire. This effect has been very marked in the cross with Eastern blood, and especially in the influence of the Arabian horse as a sire. In-and-in breeding would simply have led to degeneracy. A cross, but a judicious cross be it said, is essential to success.

On this point I may once more quote that indefatigable searcher after truth in cause and effect, Darwin.

'In the first place,' says this great authority, 'I have collected so large a body of facts, showing, in accordance with the almost universal belief of breeders, that with animals and plants a cross between different varieties, or between individuals of the same variety but of another strain, gives vigour and fertility to the offspring; and on the other hand, that close interbreeding diminishes vigour and fertility; that these facts alone incline me to believe that it is a general law of nature that no organized being fertilizes itself for a perpetuity of generations; but that a cross with another individual is occasionally—perhaps at long intervals of time—indispensable.' 'From these facts,' he further goes on to say, 'I am strongly inclined to suspect that, both in the vegetable and animal kingdoms, an occasional intercross with a distinct individual is a law of nature.'

Consanguineous intercourse has developed more imbecility in the human race than has sprung from any other cause whatever. This fact will, I think, be readily admitted. In villages and hamlets, where the poorer inhabitants have little or no opportunities of seeing any but their own relations, near or distant, marriage or other intercourse with such close relations produces the result I have described. Some thirty or forty years ago, however small the village or few the inhabitants of any hamlet, scarce one could be found in which idiots were not seen, or men and women intellectually below the general order of human beings, or, as they are in many places termed, 'daft.' Happily this disease has been somewhat arrested in such out-of-the-way places (whatever may be the result in others), since the introduction of railways, which have given the population greater facilities of mixing more with mankind, and so of lessening the number of intermarriages within limits of affinity.

I have myself collected facts on the point, from the effect of interbreeding amongst other animals. Mr. Sadler, an estimable gentleman, will be remembered as the breeder and owner of *Dangerous*, the winner of the Derby. He was a sportsman and fond of cocking, bred and kept cocks, and fought many mains with success. He used to say in-and-in breeding was a mistake, and that, for experiment sake, he tried it with his birds, and found after the second or third consanguineous cross the birds lost their natural courage, and to such an extent that some would not fight at all, and others only for a short time. Without for a moment saying that in-and-in breeding is not to a certain extent, but to a limited extent only, allowable and even wise, as I may show by example, here at least is an illustration of its failure through an abuse of the system.

One other curious fact is stated by Mr. Sadler, which, though not strictly bearing on our subject, may be set forth here as showing how strange are some of the influences affecting the dispositions of the young: 'It has been known,' he says, 'that birds bred from the purest game-fowls will, if hatched out under a common-bred fowl, partake of her nature, and only fight, like dunghill fowls, so long as victory seems assured to them.' He also told me that he once possessed a tame raven, which he induced to sit upon and hatch out some eggs of the common fowl. In consequence, the breath of the young chicks 'stank horribly,' like that of the raven which had hatched them. This, whatever may have been the explanation of it, is a very curious fact, of course; but I can only give it for what it may be worth.

Taking another analogous case from the greyhound: we know that some greyhounds lose the natural courage of their ancestors, and are of little use for coursing. For when running they will not attempt to rush through or over any obstacle that offers more than ordinary difficulty to their progress, but rather refrain from following their game. May not this be owing to consanguinity? I do not say it is, but it is very probable, because we know that crossing a bulldog with a greyhound bitch gives the offspring courage; and in the seventh generation they become as swift as others that have not been so crossed with a slower animal, partaking of all the qualities of the dam and none of the original sire (bulldog) in external appearance, yet retaining for several generations the unmistakable pluck and courage of that sire. Mr. Etwell, with whose kennels I was better acquainted than any other, bred many of his dogs so. *Eurus, Egypt*, and *Lopez* were so bred, and were good runners. I believe the then Lord Rivers adopted the same cross, and was likewise successful.

I think we may infer from such results in other animals, and with much likelihood of being right in our conclusion, that in-and-in breeding has the same effect on horses; and that the result is shown on the turf in their disinclination, or it may be lack of power, to struggle against others in a well-contested race, whether long or short. I am pretty sure that it produces irritability of temper. I do not deny that we have seen, in olden times, horses bred in close consanguinity that were good as racehorses; but in recent years the instances, as a rule, prove in-and-in breeding to be a mistake—as in one case in which a mare was covered by her own son. *Melbourne* was put to his own sister, and although the produce seemed healthy, it was a failure. Nor do I think that the examples of *Knight of the Crescent* and *Harkaway* point in favour of the practice in breeding race-horses. It did not succeed with them, nor with others similarly bred at the stud. In case of defect, crossing with different blood to good-tempered but courageous horses may have the desired result; but it is much better to avoid the evil arising from breeding from such close affinity, than to endeavour to remedy the mischief after it is done. At all events, with the examples before us, and with the decided opinion of such an one as Darwin against it, I think anyone should hesitate before trying in-and-in breeding to any great extent. In fact, every description of horse in the present day is crossed, and has had the benefit of the thoroughbred cross at one time or other, the thoroughbred alone being free from admixture with other breeds for more than a century. The thoroughbred stallion has been put to the heaviest cart or dray mare. Ponies, too, have been put to full-sized thoroughbred sires, and had produce.

That horses do transmit their peculiarities from one generation to another, in some shape or form, has, I think, been sufficiently proved. But I may proceed to show that even colour, and that a conspicuous one, can be retained in the same family for over a hundred years before all traces of it are lost. Many instances of a like nature may be found, I have no doubt, if more were necessary to establish the fact; but the one that follows is remarkable:

Crab was a grey horse, got by Alcock's *Grey Arabian* (bred probably about 1710), out of a *Basto* mare (sister to *Soreheels*), and foaled in 1722. The mare had, between 1721 and 1739, nine foals, and not one of them was the same colour as *Crab*, so he can hardly be said to be a pronounced or very decided type of his colour. *Gaudy**

---

* *Gaudy* is described in one place as a bay, in another as a grey, but in either case illustrates the fact most clearly. *Duchess*,

was foaled in 1760 by *Blank*, out of *Blossom*, by *Crab* ; she had fourteen foals, six of which were greys, and the rest bays or browns. One of the six was *Duchess*, a grey mare bred in 1777, and in 1799 she produced a grey foal called *Grey Duchess*, by *Pot*-8-*os*. She, in her turn, bred ten grey foals to six different stallions, which is a very curious fact ; for, moreover, her other ten foals were of a different colour. *The Bride*, a grey mare, foaled in 1817 out of *Grey Duchess*, bred in 1822 a grey colt *Second Sight*, and died without having a filly ; thus showing that from the year 1720, when the sire of *Crab* would most likely have been put to the stud, his own colour (grey) was transmitted through many generations to the year 1822, or over a century.

Coming nearer our own time, we see how many grey horses *Chanticleer* got, who was of the same colour, which can be traced to him now by *Strathconan*. This is the pedigree ; *Master Robert*, a grey horse, was foaled in 1811, got by *Buffer*. *Drone*, the same colour, foaled in 1823, was by *Master Robert*, and he (*Drone*) was the sire of *Whim*, dam of *Chanticleer* (in 1843), and *Souvenir* by him became the dam of *Strathconan* (in 1863), in whose stock can be traced the peculiar and uncommon colour (grey) in most of his descen-

---

for over fifty years, through *Grey Duchess*, *The Bride*, and *Second Sight*, bred foals of a gray colour.

dants, *Buchanan*, to wit. Thus from 1811 to the present day, or for about seventy-six years, the colour has been transmitted in the maternal or paternal line, and probably will be as conspicuous in the descendants of *Strathconan* and *Buchanan* in another century, as *Chanticleer's* stock has been in this, or as *Crab's* was earlier.

Again, *Grey Friar*, by *Hermit*, got his grey colour from *Madame Vestris*, a grey mare, who bred several grey foals, *Nell* being one; she did the same, and her daughter *Nan Darrell* bred *Spinster*, a roan mare, and *Perseverance*, out of her, was also the same colour, so that the grey colour on the dam's side showed much the same consistent character as in the other cases.

This peculiarity of colour-descent is seen in other varieties. The grey hairs on the flank are seen in *Birdcatcher's* stock, *Knight of St. George*, *Warlock*, and through father and son in *Partisan*, *Venison*, and *Kingston*. It is the same with brown horses, as for example, in *Blacklock*, *Voltigeur*, *Vedette*, *Galopin*, *St. Simon*, and others.

It must not be inferred, from what I have said, that I attach any importance to colour, or have a predilection for a black over a brown, or for a bay to a grey. To my thinking, it has no influence whatever over the animal's qualities. I would almost say that I would as lief breed from a cream-coloured animal with spots, did it in respect of pedigree and shape satisfy me, if I did not know that the colour, or pattern, is not to be found in well-bred horses. I have gone into the matter of colour, first because it will be expected that I should express some opinion on the point; but chiefly as illustrating the importance of my general theory, that the offspring is likely to, and in many cases actually does, inherit the qualities of the parents and remoter ancestors in a greater or less degree. We cannot, until tried, tell what temper a horse has, nor whether he be good for anything or nothing. But just as in the case of colour, we cannot tell what a foal may be till we see him, yet may form some opinion of it in the probability of his being like his parent; so in other and more important matters, if his progenitors have been good-tempered, sound, and courageous, we may reasonably expect him to be the same.

CHAPTER XII.

GENERAL PRINCIPLES OF SELECTION.

Books not much help; experience and observation the true guides —Transmission of peculiarities; unsoundness; lameness— 'Stonehenge' on the subject—Pritchard's views—Control of man over hereditary modifications in animals; Darwin—Conformation; necessity for matching sire and dam in this respect —Sir Tatton Sykes' stud, and my opinion of it.

Judgment of powers not possible without trial; instances in point—Pedigree a safeguard in selection—The breeder's difficulties—Stallion and mare must possess like features—Speed and stamina desirable on both sides—Diversity of opinion on in-and-out crossing—Why so few mares breed good horses; a theory as to the reason; examples.

The best age to put mares to the stud—Lord George Bentinck's experiments—Various illustrative instances—The rule, and exceptions to it—Early breaking-in not a deterrent to long work; management; hunters have lasted long—Work of English race-horses compared with that of American trotters.

Summary of the principles of selection.

THERE are principles in horse-breeding that cannot be gathered from books, and for proficiency in which we must rely upon the breeder's natural aptitude and faculty for acquiring experience from his own observation. By way of illustrating this, I will take an example of another kind. I hold that a good carter, one that is master of his business, though he may not be able to read or write, would yet perform his work better by mere rule-of-thumb than could an educated man who essayed for the first time to turn theory into practice.

These remarks are pertinent to the introduction of perhaps the most important part. as it is undoubtedly the most intricate, of our study; that is to say, the principles of selection. It is a subject I find it rather difficult to deal with in satisfactory sequence. I think it will be best to devote this chapter to general principles, amongst which will be included consideration of early and late maturity, and the age at which horses can best be put to the stud, with facts I have collected as to the choice of mare and stallion, irrespective of class, and of crossing the two with best effect, leaving details specially pertaining to individual classes to be examined subsequently.

I think I have proved, beyond question or doubt, that animals do transmit their peculiarities to their offspring through many generations; and that this is the case in regard to unsoundness generally, unless that defect has been created accidentally. Its detection, and the detection of incipient signs of it, must often depend upon the acuteness and diligence of the breeder himself. That unsoundness does exist in our thoroughbred stock (and from it is likely to be disseminated through all other varieties, all of them, as I have shown, owing something to the thoroughbred strain) is an undoubted fact, although one that is little thought of by some people, and by others discredited or unknown. But it does exist inherent in our horses to a great extent; and though the cause may be obscure, the fact is patent. It behoves breeders, therefore, to use every effort to minimize the evil, and in this well-directed effort I would be inclined to add the trainer should co-operate, if I were quite certain that it lay within his power to do much towards it.

A great deal of the lameness in our horses, I fear I must confess, is attributable to some cause or other not yet discovered. How far it is owing to the fault of the breeder in breeding from stock that transmits the disease, and how far it is to be laid to the charge of inexperienced or, in some cases, incompetent trainers and to bad management, is not easy to decide. In the elucidation of these particulars we have, I think, made little or no progress for many years. Some observations of 'Stonehenge' upon the subject, made years ago, are as pertinent still as they were at the time they were written:

'One chief difficulty of the trainer now,' he remarks, 'is to keep his horse sound, and, unfortunately, as disease is in most cases hereditary, and too many unsound stallions are bred from, the difficulty is yearly on the increase. Without doubt, roaring is far more common than it used to be,* and the possession of enlarged joints and back sinews is the rule instead of the exception. During the last ten years, the Derby has five times been won by an unsound animal, which the trainer was almost immediately afterwards obliged to put out of work, either because of diseased feet, or on account of a break-down; and yet few breeders think of refusing to use such horses as these.'

Now, I hardly know that trainers have had much to do with the soundness or unsoundness of horses, or that they could do much to minimize or prevent the evil. But to say more than has already been said on this subject would be only vain and useless repetition.

Naturalists tell us, with seeming probability, it is beyond dispute that animals in a state of captivity are more liable to disease than those which retain their primitive freedom. Hence it appears more than ever necessary that we should jealously guard against breeding from horses having any malformation or disease, whether hereditary or acquired. In *The Veterinary Journal*, October, 1886, appears an essay by Professor Pritchard, entitled, 'A Few Hints Respecting Breeding,' in the course of which he supports my opinion as follows:

* My own opinion on that head has been given in a previous chapter.

'This brings me to what I consider to be one of the principal points of my paper, namely, if it is necessary to be particular in the selection of dams and sires to secure size, form, bone, colour, stamina, and temperament, which I think is undeniable, how much more must it be necessary to select so as to ensure soundness of limbs and health of constitution!'

Although in breeding there is undoubtedly much uncertainty as to the ultimate result, yet the seemingly greatest difficulties may, by judicious management, be overcome to a greater or less extent, and insomuch it is unquestionably within our power to use some control over the produce of domesticated animals, and amongst them the horse. I may once more fortify my own expressed opinion by quoting from Mr. Darwin. On this subject he says: 'We may thus see that a large amount of hereditary modification is at least possible, or, what is equally or more important, we shall see how great is the power of man in accumulating by his selections successive slight variations.'

For example: with regard to the size of the legs and feet of a horse, and how he stands on them, which is one of the most essential points to be looked to, but one too often sadly neglected. If a mare, having substance and standing well on her legs, with good flat knees and large hocks, be put to a horse similarly made, the offspring will be almost certain to resemble its parents in these particulars. I do not know exactly why this should follow; for, though we might expect a lengthy horse to get a lengthy foal out of a long mare, we know *that* result is not invariable. But in this case, whatever may be the cause, the effect is undoubted.

It is not sufficient for the purpose that a stallion should possess the formation it is desired to correct in the mare, or that she should have qualities likely to improve defects in him, in order to ensure the obtaining of a foal truly formed in these respects. Both sire and dam must also be shaped alike, and have come of parents formed in the same way too, if it is desired to breed progeny as good or better than they are. I can give no better illustration of this fact than by referring to the way in which the late Sir Tatton Sykes bred his stock.

Sir Tatton had not a mare in his possession, so far as I saw, but stood well on her legs, with plenty of bone, and with good hocks and knees. He would choose stallions that were stayers, in the same way, too. Now and then he may have had one not so shaped, but he would take care to get rid of it soon, and replace it with one more suited to his well-known taste in this respect. It was a great idea in breeding, and as good as it was great, that both sire and dam should stand well on their legs, or 'stand pretty,' as the worthy baronet used to express it.

Whatever the qualities of the horses bred at

Sledmere may have been—and I have already quoted an opinion in their disparagement—one thing is certain, I think, that no one ever bred so many horses, sound in every respect, as did Sir Tatton Sykes. Of this I have had ample proof from my own knowledge of them. It would be quite as easy for others to breed animals equally sound and capable of standing work, if they would only do as he did, select sires and dams of proper shape and good constitution.

As for the opinion that no one ever bred a worse lot of horses than did Sir Tatton, I can only say that my knowledge of them does not incline me to such a view. I remember, at least, his having bred *Grey Momus*, *Black Tommy*, *The Lawyer*, *St. Giles*, *Gaspard*, *Elcho*, *Dulby*, and *Lecturer*, to say nothing of several minor winners. Of course I shall be told that he never bred a Derby winner. But he was second and third for it with *Black Tommy* and *Grey Momus*, which is a great deal more than many other breeders can boast of. But for some of these it is of no consequence how animals are shaped, or how their legs are put on, if only they have four of them, and the regular number on each side. They think that by putting a large bony horse to a little twisted-leg, light-limbed mare, they are sure of obtaining a foal with good bone, standing in the most perfect way imaginable. They are surprised by the natural result, unexpected by them, of such an ill-assorted union; yet, if the offspring happens to be pretty well bred, they will keep on trying to dispose of it at any and every yearling sale, which generally ends in its being sold to some unlucky person for a 'pony' or a twenty-pound note.

It is a truism, I may observe, a fact well known to all owners, trainers, and breeders of race-horses, that some horses are faster than others, that some have unquestionably great staying power, and that a fewer number are possessed of both speed and endurance. Nevertheless, while this is admittedly the case, there is no method by which the different qualities of the animals can be discovered except by trying them, and so ascertaining what their several powers really are. The possession of speed or stamina on the one hand, or the lack of either or both on the other, cannot be judged from the conformation merely, which may be closely alike in horses of very different qualities. Some persons, whose opinions are entitled to respect, tell us it is a sound and vigorous constitution that enables one horse to stay better than another, irrespective of shape and make; and that it is a faint heart which prevents others doing as well—views I am disposed to endorse.

It is only too easy to say of a horse: 'He has a good game head, and is sure to stay;' or, 'His chest is deep, with plenty of room for a heart as large as a bullock's;' or, 'He is too long;' or, 'He is

GENERAL PRINCIPLES OF SELECTION. 157

too weak to get over a distance of ground under any circumstances.' In such cases, judging from external appearance only, you may be right, or you may be wrong, though, of course, a man well accustomed to such matters is more likely to be right than one of less experience. Still, it is little more than guess-work, and often of no value.

We have seen, for example, the weakest light mares, such as *Mendicant* and *Fraulein*, and many others I could mention, which were able to stay any distance, and to carry weight. Yet, if before we had seen them do it we had ventured an opinion, it would have been something of this sort: 'They may get half a mile and win a race with a light-weight upon them, but they can never do any good beyond that distance or under a heavy-weight, and are useless for breeding purposes.' I think, therefore, if we are to judge of the merits of any horse or mare before it has been proved, we must have recourse to other methods, or we shall probably be mistaken. What would be a real safeguard, under which disappointment should prove the exception rather than the rule, would be to make sure that the animal we were judging of for selection had really superior breeding, coming from a long line of good ancestors on both sides, whose collateral branches had shown the same excellence, and, consequently, with only such intervals in close breeding as would be sufficient to prevent the evils of consanguinity.

The task of selection is not completed when, first, a mare has been chosen that in blood and all other qualifications appears satisfactory; nor yet when, second, a stallion has been secured possessed of every needful requisite. It is quite as imperative, if the breeder would be successful, and almost more difficult of accomplishment, that the sire and dam should be so suited in their respective qualities as to render it likely that their progeny shall be characterized by the best points of both. To the achievement of this, as of other technical details, there is no royal road, nor ever will be. So many are the difficulties which the breeder has to overcome, that one is almost tempted to believe there must be no small amount of luck in producing the ideal foal. No sooner have you found a sire and dam with whose pedigrees you are charmed, than you discover unsuitability in the conformation of one or both of them. Either may be too large, or too small; or one may not have such points as are needed to counter-balance defects in the other; and thus no chance appears of breeding anything above mediocrity, if even as much as that.

I have previously noticed the tendency that exists among almost all kinds of animals to revert to parent stock, namely, to produce offspring possessed of characteristics inherited from remote ancestors. This is a most important fact, and one it is necessary to keep in mind when breeding from different strains of blood—'crossing,' as it is technically termed.

To produce the greatest number of good horses from any selected couple, I believe it is necessary that both of these, the original sire and dam, should be possessed of speed and of stamina. I know it is always difficult to secure these advantages on both sides; yet it undoubtedly may be done in many cases, and always should be whenever possible. For example, *Touchstone* and *Crucifix*, both known stayers and speedy animals, produced *Surplice*. From the same sire and *Beeswing*, also a good stayer, came *Newminster*. From *Weatherbit* and *Mendicant* came *Beadsman*, and from him again *Blue Gown* and *Rosicrucian*. *Voltigeur* was by *Voltaire*, a son of *Blacklock* out of *Martha Lynn;* and from the same blood on the paternal side came *Vedette*, *Galopin*, and *St. Simon*, all of whom had speed and could stay, so far as we know. These examples might be added to, but I think should be sufficient to show how desirable it is to breed only from the best horses, both sire and dam, having due regard to speed and stamina. I shall revert to this topic again when treating of the thoroughbred pedigree.

That there should be a great diversity of opinion amongst breeders, as to the most approved method of obtaining the best stock, is not to be wondered at. One man believes in large mares, and another in small ones, whilst others say, and I think rightly, that a medium-sized mare, as well as stallion, will excel in producing stock for the turf. Then, again, the breed of one animal is often as much disliked by this owner as it is admired by that.

The same difference of opinion exists in relation to crossing, and scarcely two professed judges are in agreement on all points connected with it. Mr. Smith, who wrote on breeding for the turf many years ago, said : 'Once in and once out ;' that is, once with the same strain, and, in the next generation, with a different one. Breeders of more recent date have adopted the principle of 'twice in and once out,' and there are some who advocate in-breeding to the closest degrees of consanguinity. Many good judges hold that no relations nearer than the third or fourth cross should be permitted to come together. I have already given my opinion on in-and-in breeding, and need not now say more than that I do not approve of putting near relatives together.

Another problem is why so few mares breed good horses. In my opinion, the solution will always remain hidden among the secrets of Nature. One would be inclined to suppose that if a good mare be put to a stallion equal in quality to her, the result must surely be excellent. But this is too often not the case, the son turning out inferior to both his parents, and not worth his saddle-girths as a racer. Again, it sometimes happens that a horse unexpectedly turns out far better than could have been looked for from his pedigree. Such cases are quite inexplicable. Generally, as I have observed, or, indeed, my present task would have been superfluous, there are principles which, if followed in the selection of stock, will lead to success. Yet there are, as above stated, exceptions to the rule, astonishing because unaccountable. I cannot do better, perhaps, than supplement my own opinions, already expressed on the point in one or two places, with those of a gentleman who is one of the best judges of a horse in England, and, I think I am right in adding, the most successful breeder.

'Why so few mares breed anything worth breeding,' he says, 'is because cross-bred stallions are put to cross-bred mares. Should I have a cross-bred mare, got from stock cross-bred for five or more generations back, I would wish to put to her a stallion of at least two strains better—that is, one whose pedigree shows that the best blood in him has been obtained from more than one ancestor of the same origin, and this either for speed or stamina, as the case may be. For instance, if I bred from a speedy mare, I should look for her staying crosses, and try to get a stallion having two of them in him, and of as near generation as I could find. Especially ought this to be attended to if the stallion's speed in training were better than the mare's. For it is an entire mistake to suppose that the best stayers are got by stayers. No breeder could ever prove such a dictum to be a certainty.'

I have collected a few examples illustrative of this very interesting point, which bear out the remarks I have just quoted in a considerable measure.

DONCASTER,

By *Stockwell*, out of *Marigold*.
Marigold by *Teddington*, out of sister to *Singapore*.
Sister to *Singapore* by *Rattan*, out of a *Melbourne* mare.

Here we see distinctly the two crosses for stamina and speed. Through *Stockwell* and *Melbourne* he got his staying powers; his speed was derived from his great-great-grand-dam *Vulture*, through *Teddington*.

BEND OR,

By *Doncaster*, out of *Rouge Rose*.
Rouge Rose by *Thormanby*.

Here what staying-power *Bend Or* has is traceable to *Thormanby*, who derived it from his mother, *Alice Hawthorne*.

ORMONDE,

By *Bend Or*, out of *Lily Agnes*.
Lily Agnes by *Macaroni*, out of *Polly Agnes*.
Polly Agnes by *The Cure*, out of *Miss Agnes*.
Miss Agnes by *Birdcatcher*.

Ormonde is a horse bred more for speed than endurance on his mother's side, through *Macaroni*,

The Cure, and Birdcatcher. Whatever staying powers he may have must therefore be looked for on the paternal side, though at a great distance. But whether he will stay better, or not so well as his great grandsire, Thormanby, remains to be proved, as he has never yet run a greater distance than a mile and three-quarters; though from his performance over that distance it may be inferred that he has inherited Thormanby's stamina.*

PETRONEL,
By Musket, out of Crytheia.
Crytheia by Hesperus.

In this case both dam and grandsire were short runners; yet the former, when mated with an undoubted stayer, Musket, produced a foal as good in this respect as his sire.

I will now refer to some matters connected with breeding; the first of which is concerning the age at which mares are best fitted for stud purposes, and the question whether much racing is or is not prejudicial to the brood-mare. Here, as in other cases, it will be best to seek guidance from actual facts, and rely on them, rather than on the fertile resources of imaginative speculation.

Lord George Bentinck was the most indefatigable racing man of his time. In breeding, as in

* Since this was written Ormonde has been reported to have turned out a roarer, as I have elsewhere noticed.

racing, he tried everything that he thought likely to prove a success. He put *Monstrosity* to the horse as a two-year-old, and, as her son, *Ugly Buck*, was a success, he carried the idea still further by putting *Experiment* to the horse a year younger—in fact, before she was twelve months old. But in this case Nature resented the liberty taken with her general laws, and the result was total failure; for, though the mare had a foal, it only lived a few hours; she had one after, slipped the following year, and then was sold. *Monstrosity* bred *Ugly Buck* in her third year, but had nothing so good subsequently. *Pasquinade* was also put to the horse at two years old, and bred *Libel*, her first foal. *Paradigm* was the dam of *Lord Lyon*, *Achievement*, and many other good horses afterwards; though she did not run at three years old, was not covered then until the following year, and continued to breed with great success up to her twentieth year. *South Down*, the dam of *Alarm*, had her first foal at four years old, and continued to breed till she was twenty-four. *Cavatina*, *Trumpeter's* dam, was the same age when she was sent to the stud. So was *Maid of Palmyra*, the dam of *Viridis*, and *Mandragora*, from the last of whom came *Miner* and *Apology*, winner of the St. Leger. *Polly Agnes* was put to the horse at four years old, and bred till she was sixteen. The last I shall mention was *Bas Bleu*, who was covered at four years old and bred *Blue Gown*, had her last foal in 1880, being then twenty-two, and was destroyed at the age of twenty-five.

Sir Tatton Sykes usually put his mares to the horse at five or six years old, and he certainly bred successfully. Although *Crucifix* was not put to the stud until five, I see no reason why mares that have been kept well, as most thoroughbred mares are, may not be covered earlier—as I have said, at three or four years old. But I should prefer not having a mare covered, if she had been racing until she was four. At that age I think the procreative power is fully developed; many mares, at from four to six, have produced stock as good, if not better, than any they have subsequently bred. *Libel, Miner, Ugly Buck*, and many others, are examples of this. I consider young or middle-aged mares the best breeders on the whole; for, though a few will breed till they are twenty-four and over, I can call to mind none such that ever produced a good stayer much after twenty, either by a young or an old stallion.

If that be the rule, however, it is not without some exceptions, as I know. *Pocahontas*, when twenty-five, bred *Araucaria*, a winner, by *Ambrose*, who was then twelve years old. *The Roe* was as old when she bred *The Cob*—a good fair horse, and perhaps the best she had—by *Lord Ronald*, who was eighteen when he covered her. But much dependence can not beplaced on isolated cases such as these, and it is for that reason I have elsewhere explained, I hope succinctly, some of the curiosities of breeding.

Some people think that, after racing, mares should have a year's rest before being covered, but in my opinion this is not absolutely necessary. *Crucifix* and *Paradigm* were so treated, but *Flying Duchess*, and *Pocahontas*, on the other hand, were put to the horse the year following that in which they concluded their racing, as is the usual practice. The former was four and the latter five years old at the time, and both of them bred good winners, including winners of the Derby and St. Leger.

Facts are stubborn things, and they certainly seem to contradict the popular notion, that the older horses are before they run, the longer they continue on the turf, and the greater the number of times they can run—though there is no rule without some exceptions to it. Now, *Eclipse* was five years old before he ran, and yet he was only two years on the turf, whilst, as I have elsewhere shown, *Touchstone*, for one, commenced earlier and lasted longer, as hundreds of other horses have done. *Eclipse* covered at Clay Hill, near Epsom, at fifty, thirty, and then twenty-five guineas each mare, by which it appears that his reputation as a sire did not increase with his age. He died in 1789.

'However,' says Mr. Woodruff, speaking of trotters, 'horses do not attain their best form until they are seven or eight years old, and then they will last until they are eighteen or twenty, and even longer. One that is worked much, or run several times, at four or five years old, will be worn out in two or three years.' We know hunters, in this country at least, if not broken in until three years old, and then not ridden hard until they are six or seven, will last until they are twenty; indeed, some have been known to carry their master during as many consecutive seasons. These, however, are all mares or geldings.

We have a few examples of race-horses that have been trained early, and yet lasted until they were fourteen or sixteen: such as *Lilian, Historian*, with *Clothworker, Reindeer*, and *Oxonian* as geldings; but there are not many such. Others leave the turf at the end of their second or third season; not a few are sold at two years old, but that is on account of their being good for nothing as racers.

It should be remarked that English race-horses of the present day, what with their training, running, and frequent trials, do as much work in one year as the American trotters would in several, which might account for their earlier decay. Yet none of the American racers, which are supposed to be older when broken, trained, and run, are as good as our horses, or last longer on the turf; at least, none that we have seen here have equalled our own. But to this fact I have elsewhere drawn the reader's attention.

Mr. Hiram Woodruff's account of how the American trotter is trained pretty much agrees with the system adopted by many of our old trainers of race-horses. Whoever is interested in the subject will find in his book much that is amusing or enlightening. *Sloe*, after winning several races, retired from the turf unbeaten; and others have done the same. *Wapiti*, after winning all the races she ran in at Goodwood, broke down, and never started again.

To sum up briefly what I have said: if it were possible to reduce all the principles of selection to one or two maxims, I would say, 'Consider carefully the external form of the mare, the relation of her different parts to each other, her capabilities, so far as known, above all, her breeding and that of her ancestors; then select a stallion on the same careful system, and you may expect, and most likely will have, a foal possessing in many ways the desirable qualities of its parents, and perhaps surpassing them in speed and endurance.'

In concluding these general remarks on the principles of selection for the improvement of the thoroughbred horse, whether the purpose be to breed for the turf, or useful animals for other purposes, I would emphatically urge that on no consideration whatever should we breed from imported horses or mares of any description. Our horses are already too much mixed with foreign strains, of whose breed, constitution, temper, or soundness we absolutely know nothing, or from what unhealthy or degenerated stock they may have come.

CHAPTER XIII.

SELECTION OF THE MARE.

Selection involves complex study—The details to be considered—Examples drawn from thoroughbreds are applicable in the case of any other class.

Breeding—Choose either distinguished or young mares—Relationship to good horses my great point—Mares of good blood justify their selection on that ground; examples—Mr. Wreford's experiences—A good performer, if of bad stock, is generally disappointing—Some 'good-looking failures'; no reason assignable—The element of chance—Three lucky purchases: the *Cervantes* mare, *Grace Darling, Octaviana*—Three unfortunate purchases: *Marie Stuart, Mandragora, Agility*—The lessons these cases teach us.

Shape and size—My preference—Small mares of strong frame the best—*Glee ;* other examples.

Action—Mares with bad action often failures—Mares with good action produce clever horses—Examples—This point should be studied—Action may be developed, but originates from breeding—Curious instance in a dog.

Constitution—Soundness and absence of all disease essential—How to judge of this—*Alice Hawthorne* and *Beeswing*—The pedigree a test.

Speed and stamina—Fast mares preferable for the stud—Summary of all the points to be considered.

In the choice of a mare for stud purposes, so many details arise for consideration, that the feat itself seems to be the effect of somewhat complex study. But if, in selecting, we proceed in an orderly method, we shall readily arrive at a definite conclusion on each point as it comes before us, and finally on all put together. We must examine in detail the subject of breeding, also shape, size, action, constitution, capacity for speed and stamina, before we can reach our ideal.

The examples, from which I shall here illustrate my opinions, must be, almost of necessity, deduced from my experience with thoroughbred stock; but they should not be lightly passed over on that account by breeders of other descriptions of horses. Everyone who breeds horses of any kind does so in the hope that he will breed the best of that kind; and with the sanguine this hope amounts in many cases to thorough conviction that such will indeed be the result; for all breeders know well, or ought to know, that, in the end, only the best stock will really pay them. Very few, however, are sufficiently alive to the fact that, in order to effect their aims in this respect, it is necessary to commence with stock calculated to produce the desired result. Wherefore, the study of rules which should be applied to the choice of thoroughbred mares, must teach a great deal of the first and most important principles to be applied in the selection of mares of any other kind.

First, as to breeding, we know the truth of the

aphorism that 'Like begets like.' Still, this is a principle which must not be relied on too implicitly. Reason and common-sense must guide our judgment as to what we shall select or reject. Yet the maxim is not to be disregarded; therefore, let us start with the best-bred mares we can get hold of, not grudging the cost where it can be reasonably expected that the investment will pay.

If I could afford it, my mares should be those which had been the best runners on the turf; or such as had bred winners; or again, such as had retired early from their racing career owing to accidental circumstances, yet giving promise of better things, like *Paradigm*, for instance, the dam of *Lord Lyon* and *Achievement*, and others. This mare only ran twice in public, being just beaten by *Lord of the Isles* in her first race. If such mares were not to be had, except at extravagant prices, I would rather fall back upon young well-bred ones that had only just been put to the stud, or those with a first foal at heel, than buy old mares that had been breeding some time without favourably distinguishing themselves in their offspring.

With regard to blood, I think that relationship to good performers is even of greater importance than extraordinary excellence in the mare herself. If she shall have proved herself a great performer, then of course she will be a more valuable addition to the stud; but, should she prove a failure there, she would be a heavy loss for the breeder. In the former case she will not entail such expense, and yet will possess all the advantages which can be desired. For instance, if a mare should come under my notice which had bred several notable horses, and then a filly of no use as a racer, albeit own sister to, perhaps, several of the best horses of the day; and if this filly were of fair size, stood well on her legs, and seemed otherwise well fitted for stud purposes, I would go no further in search of a brood-mare, but would buy her at once, if I could get her at a reasonable price. For experience has taught me that such an animal is likely to produce runners. In fact, a mare descended from stock that has been successful on the turf, and at the stud, is morally bound, if I may so phrase it, to justify your selection of her, unless, of course, there be anything decidedly objectionable in her shape and make.

Cases in point may be cited from the experience of Mr. Wreford, of Gratton, North Devon. He bought *Margellina* for five-and-twenty pounds. In herself she was not worth even a tithe of that small amount as a racer; but she was own sister to *Memnon*, winner of the St. Leger. And what was his experience with her? Why, she scarcely ever bred him an animal that could not run, producing *Warden, Westonian, Wahab, Wedding-Day,* and *West Countryman*, all throwing back to her better relations. From the experience of the same gentleman we may learn how ruinous at the stud may be the best mare in public form if she come from a bad stock. *Mouche* was bought by Mr. Wreford for seven hundred guineas, after she had run second to *Variation* in the Oaks. But, though put to many of the best stallions, she never bred anything but *Worthless*, a good fair horse, however, despite his name.

In short, there is, in my opinion, little advantage in selecting a mare that has run successfully if she comes from a bad stock—that is to say, if her brothers, sisters, dam, or grand-dam, or all or any of them, have been bad. I could multiply proofs on proofs in support of this assertion, showing how mares, good in themselves, have yet cast back to their degenerate relations. Nor does it do to be always quite certain of success when you have got a mare with a splendid pedigree, and as beautiful to look at as she is good in blood. For one thing, such handsome creatures are always costly to buy, apart from the fact that there is absolutely no certainty as to what the qualities of their progeny will be. I will give a few examples of what I may term 'good-looking failures.'

Largesse was a fine bay mare, standing about 15 hands 2 inches. She was by *Pyrrhus the First*, winner of the Derby in 1846, out of *Mendicant*, winner of the Oaks the same year. Could better parentage be wished for? Yet she was put to seven different stallions, and bred a foal to each of them, but not one that turned out worth its keep. Again, *Rapidan*, by *Beadsman*, Derby winner in 1851, out of *Miami*, winner of the Oaks in 1847, had eleven indifferent or inferior foals, the best of which, I think, was *Lorna Doone*, by *Scottish Chief*. Then, too, *Delilah*, by *Touchstone*, out of *Plot*, by *Pantaloon* (than which breeding no better could be desired), bred seventeen foals to different stallions, and all of them proved worthless as racers. *Schism* was another such case. She was by *Surplice* out of *Latitude*, sister to *Elis* by *Langar*, as fine-bred animals as any in the 'Stud-Book.' She was a good mare, though not quite first-class, standing 16 hands 1 inch high. But, though put to many good stallions, she never bred one foal that was worth a guinea, though several of them were good-looking enough.

The reader may be inclined to think these experiences rather bewildering at first sight, when regarded as lessons, especially since I am unable to account for the failures on any sound theory. In the case of *Schism*, however, I think we may conclude that she threw back to *Elis*, brother of her dam, who got few if any good foals. In the other cases the mares may not have been mated with stallions properly suited to them, or, from some undiscoverable cause, they may have been faulty by nature. But there is one inference to be drawn from these cases; at least, I can only come to the conclusion that a given amount of uncertainty remains, even after a mare has been secured that in looks and pedigree is everything that could be wished for. It is obvious, therefore, that one should hesitate before paying such large sums as frequently are paid for 'good-looking ones.' It would surely be wiser to first ascertain if better value cannot be got for the money by purchasing mares not quite so handsome, but possessed of all other desired qualities, leaving the 'beauties' to such as breed more for fame and in accordance with their fancy than for profit, and who are able to afford the risk of expensive failures.

That 'luck' may be with or against the breeder, after all is said and done, is not to be gainsaid. Here is an interesting account of six mares, of what led to their purchase, of their respective cost, and subsequent value at the stud. The story of the first of these runs as follows:

Sir Richard Constable called upon Mr. Henry Robinson, the elder, of Carnaby, prior to a sale of thoroughbred stock which was to take place in the neighbourhood, and asked him if he intended buying any of the mares. Mr. Robinson replied in the affirmative, and named the mare he thought of bidding for.

'I am sorry for that,' said Sir Richard, 'as I should like to have bought her myself.'

Thereupon Mr. Robinson very generously responded:

'If you don't want the *Cervantes* mare, I will buy her instead, and will not oppose you in the purchase of the other.'

'Thank you,' said Sir Richard; 'then I will not bid against you either.'

In consequence of this arrangement Mr. Robinson became possessed of the *Cervantes* mare—which had been covered by *Humphrey Clinker*, and afterwards bred *Melbourne*—for the moderate sum of ten guineas, without having had the original intention of buying her.

The next case is that of *Grace Darling*. She had produced nothing worthy of notice before her sale, except the foal then at her side, who was afterwards known as the redoubtable *Hero*. She and this foal were put up for sale at Lansdown on one of the days of the Bath Races, and they were knocked down to Mr. Whitwick of Codford, a sporting parson, for the sum of fourteen guineas. Immediately afterwards he resold them to Mr. John Powney for an advance of two sovereigns on the first price. I suspect neither of these gentlemen had ever before possessed a thoroughbred mare or horse, and hence were hardly likely to have been the best of judges, their purchase of *Grace Darling* being more or less fortuitous.

In another instance there was a very similar appearance of luck. *Octaviana* had been in Lord Chesterfield's stud for some years, and was seventeen, having bred nothing of consequence. Lord George Bentinck bought her for sixty-five guineas, together with the foal then running at her side, which was afterwards known under the name of *Crucifix*. *Octaviana* subsequently produced nothing of any note, and died in 1841.

These are examples of the curiosities of breeding, and what I am going on to tell are equally so, though not quite of the same kind. They relate to three mares which were failures, yet which changed hands at prices larger, probably, than any given before or since for brood-mares.

The first of these was *Marie Stuart*. She was sold, I believe, by private contract, for the sum of £8,000. But she never bred a winner either before or after that sale—though I am aware she ran again after it—and indeed only had one foal living in all her career at the stud.

The other two were *Mandragora*, and her daughter *Agility*, which the late Mr. Gee bought at Doncaster after the decease of Mr. King. If I remember rightly—for I must quote this from memory—the price he gave was 3,000 guineas for each of them. At any rate, neither of them ever bred him a winner of a plate of the smallest value. At that time Mr. Gee was forming his stud at Dewhurst. He was a person of but little, if of any, experience in such matters; and neither, I believe, was Mr. W. S. Crawford (the purchaser of *Marie Stuart*) considered a capable judge of breeding.

Anyone with a competent knowledge of the subject would never have expected to obtain anything worth having from these mares. In the first case, that of *Marie Stuart*, her dam was *Morgan la Fay*, who had no living produce for seven years, and was barren for five. What, then, could have been anticipated from her daughter but sterility or disease? In fact, *Marie Stuart* was barren for nine years, or only produced dead foals; up to the time of this writing she has given birth to but one alive —*Abbey Craig*.

Inexperience might have been pardoned for hoping more from *Mandragora*, however, since she was the dam of *Apology*. Yet she had borne several dead foals, and had been barren, and from the same cause she was useless at the stud for eleven years. What could have been looked for from the daughter of such a mare but an equal degree of badness, and, indeed, utter failure! And this was just what did ensue; for *Agility* was barren in 1874 and 1876, and died at the age of eleven from disease, which she doubtless inherited from her mother.

The shape—with which it will be convenient to include also the size—is the next point to be considered in forecasting the possibilities as to how a mare's produce will turn out. For my own part, I prefer small mares, low, long, and compact, standing 15 hands to 15 and 1 or 2 inches. I do not mind if they are a little short, provided they have substance; for I consider frame to be of more importance than height. A mare with plenty of strength, though short in the leg and not over 15 hands, would be far preferable to a taller, leggy animal with a weak frame.

We have a good example of a small mare in *Glee*, who was but very little over 15 hands high, but who had plenty of power. She was by *Touchstone* out of *Harmony*, by *Reveller* out of an *Orville* mare, and thus had good staying blood in her. She bred *Jerry Kent*, *Promised Land*, and *Happy Land*, all by *Jericho*.

I know enough of small mares to affirm that they can and do breed good horses, generally larger than themselves, and the last more particularly when their ancestors were of good size. The fact, being well proved, should be made of service in breeding for the turf, or for the stud afterwards. Among other examples we have *Maid of Palmyra*, dam of *Viridis*; *Flying Duchess*, dam of *Galopin*; *Seclusion*, dam of *Hermit*; *Defenceless*, dam of *Caractacus*; *Cobweb*, dam of *Bay Middleton*; *Alcestis*, dam of *Devotion*, the dam of *Thebais; Bravery* (a pony), dam of *Rupee; Salamanca*, dam of *Pero Gomez*, winner of the St. Leger. Not one of these mares was much over 15 hands, if at all; and instances of even smaller ones could be supplied. But these should be sufficient to show that little mares are the best to breed from, provided they have substance, stand well on their legs, and have good hocks and knees.

Action has also not a little to do with the general merits of a race-horse. I should feel very much indisposed to breed from a clumsy walker, and I think all brood-mares should be well qualified in this respect. *Wapiti* and *Virago* were both of them bad in action, and neither ever bred anything of note except *Thalestris* and *Wilderness*, which were but moderate. They were large, heavy mares, although two of the best in their day, and they were well-bred. On the other hand, *Maid of Palmyra*, the dam of *Viridis*, was a weak little mare with small limbs, but she had good action as a racer. She could not go fast, however, neither could she stay over a distance, for which reasons I parted with her, accepting an offer of £50 for her from the late Mr. Blenkiron, and deeming that price to be quite double all she was worth. Yet she had good action, as I have said, and she bred *Verdant*, *Viridis*, about the best mare of her year, and the dam of *Springfield*, probably the best horse we have had for years, and one likely to become fashionable and prove as good at the stud as he has on the turf. Now, I attribute these successes of *Maid of Palmyra* not only to her blood, which was unquestionable, but also to her action, which was superb.

In *Seclusion*, too, we have an example of a small mare, but a compact, strong-framed one, with the best possible action, producing good horses. She was the dam of *Hermit*, winner of the Derby and other great races, who has proved himself one of the best stallions for getting speedy animals that was ever known, perhaps. *Flying Duchess*, the dam of *Galopin*, also a Derby winner, stood about 15 hands 1 inch, and was very like *Seclusion*, except that she was weaker and rather longer; but both of them were good-tempered and showed splendid action. It is clear to me, then, that in choosing brood-mares their action must be studied, since on this point their success or failure at the stud will in some measure depend. Indeed, I would pay as much time and attention to a mare's manner of walking as I should devote to the inspection of a thousand-guinea yearling before buying him.

That good action comes from breeding I have no doubt; but, as will be mentioned again when I come to speak of the stud-farm, it may be developed, on the one hand, by the provision of ample room for exercise, or it may be suppressed, on the other, by the lack of that essential in confined quarters.

But that good action originates from breeding, I have no doubt, as I have said. While on this topic, I may mention an analogous instance in the case of a terrier dog, which came under the notice of a friend of mine.

A gentleman possessed a very nice fox-terrier, of which he was very proud, and which he prized greatly, on account of the 'perfect purity' of its breed. The dog was shown to my friend, and this point in its favour was especially adverted to, of course. 'Well,' said he, after carefully observing the dog, 'I never yet saw a well-bred terrier with such action as that. Why, he moves like a high-stepping carriage-horse!' Further and closer inquiries into the animal's pedigree were made, and it turned out, curiously enough, that one of its ancestors, generations back, had been crossed with an Italian greyhound. The terrier, therefore, had inherited from this remote ancestor the peculiarly graceful high action of his kind.

If this should obtain with dogs, I might be tempted to ask why it should not equally obtain with horses, were I not well assured of the fact that such is actually the case. I only recorded this curious instance in further corroboration of it. Care should always be taken in the selection of the mare, and of the stallion likewise, as to the possession of good action, not only in themselves, but also in their progenitors.

Next we come to the important subject of constitution. It is not only necessary that a mare should be well-bred and well-shaped, that she should be good-tempered and display good action; she must also have a sound constitution. I take it that a mare which breeds unhealthy or dead foals is probably diseased in herself, though to all appearance she may be as healthy as any. It behoves us, therefore, to select only such mares as have given some proof of their soundness; and this may be arrived at in two ways.

First, and this is usually the surest way, the soundness of a mare may be deduced from the fact of her having run a reasonable number of times, and having retired from the turf as sound as when she commenced racing. Noticeable examples of this were seen in *Alice Hawthorn* and *Beeswing*, neither of which ever produced a dead foal. This would be one safeguard. Another—and, after all, I do not know if it be not the best—is not to have any mare coming from a bad stock on the side of either parent, or of any relations further removed. And though speaking now of the mare in particular, I need scarcely say that exactly the same precautions as to soundness should be observed in selecting the stallion also.

Lastly, in dealing with the qualifications necessary in selecting our mare, we come to the subject of speed. In regard to this, I may say briefly that, of two mares equal in all points but this, I should prefer to breed from the speedier. *Vulture*, the dam of *Orlando*, was a very fast mare; her son inherited the good quality, and was a Derby winner, afterwards begetting *Teddington*. Again, there were *Seclusion* and *Flying Duchess*, both fast, though neither could stay, but each the dam of a Derby winner. Writing to me a short time ago, a very successful and large breeder said, 'I like speedy mares; in fact, I never did any good in breeding from slow ones.'

To sum up: I should choose a mare standing 15 hands and 1 or 2 inches in height, having a small head and a short neck, with good shoulders, back, and hindquarters, with broad hips and deep fore-ribs, with satisfactory hocks, knees, and feet, and standing well upon her legs. On the other hand, I would not breed from a mare which had calf-knees, or was pigeon-toed, or stood upright in the ankles, or had curby hocks. I am not fond of mares with straight hocks either, as they generally breed short runners, though now and then stayers; but the feature is an unsightly one, at the best. Both *The Hero* and *Oxonian* were formed in that way; the former could stay, and the latter could not. However, straight hocks generally indicate speed only. I would also make sure that the breed of my mare was unexceptionable, whether she had been good, bad, or indifferent as a racer.

For nothing would induce me to breed from a common-bred mare, however good-looking, and however good as a racer she might be. I hold this matter of blood paramount to all other considerations, deeming it by far the most essential particular which it behoves the breeder to attend to in order to ensure success.

CHAPTER XIV.

SELECTION OF THE STALLION.

Health a first essential—Difference in cases of mare and stallion—Power of the stallion to transmit diseases; case from *The Veterinary Journal*—The best horses, of pure strain and free from defects, should alone be chosen—We keep too many inferior animals.

Breeding a point of great importance—Pedigree must be closely scanned—Temperament and capabilities of ancestors to be studied for several generations.

Size and shape—Middling-sized horses most successful—*Eclipse*, *Touchstone*, and many other examples—Failures of large horses; *Prince Charlie* and others—*Stockwell* unique in this respect—Result of comparison—Shape, another essential—Only one good form—Description of my ideal.

Effect of using inferior stallions—Subsequent offspring by other horses affected—Cecil's opinion—One cross sullies a mare for life—Overwhelming evidence in proof of this—Sir Gore Ouseley's cross between a mare and zebra; result—Chestnut mare and quagga; result—Darwin on the subject—Necessity of precaution therefore.

HAVING in my last chapter described the qualifications of the mare likely to breed the best horses if properly mated, I will now proceed to set forth those which, in my opinion, are required in the stallion. In doing so, we must remember that nothing is too insignificant to be noted which really bears upon the subject, when perfection is the result desired.

I will first, then, observe that no competent judge would think of denying for a moment that good health is as essential a point in the sire as in the dam. But there is this difference. It does not matter, with the sire, what is his state of health *after* the mare has been served, provided always that he is and has been perfectly fit and well at and before the time of his admission to the mare. With her it is different; she requires not only to be healthy at the time of conception, but it is necessary for her to remain so till the foal is born and weaned; for it is only under such circumstances that she is capable of producing good, sound, and healthy offspring.

A curious account appears in *The Veterinary Journal* for October, 1886, which shows that the stallion has, if nothing else, power to transmit diseases to his offspring. I quote as follows: 'A mare bred six foals, all of which had the same action. Further, the sire of one of the foals had navicular disease, and out of the six this was the only unsound one, and this had navicular disease.' Surely this goes to prove that the sire may transmit his infirmities to the offspring as readily as he may any other features—good, bad, or indifferent.

Thus soundness and freedom from disease, inherited or acquired, both in our stallion and in his ancestors, must be looked for as keenly as in the case of the mare. In short, what I have said in reference to the latter applies, with the one reservation, in great measure, equally to the horse.

Therefore, I say, breed only from the best horses, of whatever description they may be; and from those only which you know, of your own knowledge, to be of a pure strain and free from defects in their progenitors, near or distant. By so doing you will not only benefit yourself, but the country. In sum, I think that our chief error as a nation in this respect—and it is a matter that I only now glance at, as I shall have something more to say on it later—lies in keeping both too many stallions and too many mares. Many of them are, either from constitutional or hereditary infirmities, or from lack of shape, manners, or capabilities, quite unfit for breeding purposes, inasmuch as their offspring will only too surely inherit one or other of these defects in an increasing ratio, and become more useless even than their predecessors have been.

The primary essential to be taken into consideration in the selection of the stallion is his breeding; for although a thorough knowledge of the pedigree of a stallion will not of itself enable one to choose a horse suited to the purpose, it is yet one of the grand essentials for the purpose. This qualification wanting, nothing but disappointment can be expected; therefore, as the groundwork on which to raise your superstructure, study well his pedigree before you select any horse. Some things may be dispensed with, but not this.

I may observe that I have already said so much on this subject when speaking of what the pedigrees of mares should be, that it will hardly be necessary for me to add more than that the breeding of the stallion on both sides should be studied most carefully to the third and fourth generations. For to the breeding of these earlier progenitors as much importance should be attached as to that of the sire and dam themselves; and not only to the blood, but to the individual temperaments and capabilities as well, which should be such as are fitting to mate them with superior mares. From neglecting any of these principles, disappointment too often ensues. Should any defective strains appear in the stallion, however remote, it would be better to discard such an animal than attempt to breed horses in which you may almost anticipate reproduction of the faults of their ancestors.

These principles as to breeding have a general application, and do not require to be enforced by examples. But in giving my view of the required shape and make of the stallion, I shall adduce not only reasons, but facts, for liking this one or for rejecting the other. As to size, I may start by observing that the stallions which have been the most successful at the stud for many years are those of a middling size, or about 15 hands 1 inch to 15 hands 2 inches or 3 inches high. *Eclipse* was said to be of the former height, and though, perhaps, he might not be thought much of in these days, he was, in his own, better than many horses that were said to be much bigger. Other instances of success in moderately sized horses we find in *Touchstone, Orlando, Sir Hercules, Newminster,* and *Hermit.* These were, most of them, comparatively speaking, of small size. Again, *Venison,* when in training, was barely 15 hands 1 inch, though he grew afterwards an inch or so; but he was still a pony by the side of such horses as *Bay Middleton* and *Elis.* Both these comparative giants were better race-horses and as well bred, and yet neither of them was successful at the stud. In fact, if we consider the number and qualities of the mares that both of them had, and the years they were at the stud, they may fairly be set down as failures. *Galopin* and *Vedette,* his father, were not bigger; neither were *Kingston, King of Trumps, Defence, Sweetmeat, Macaroni,* or *Weatherbit;* and if we add *Rifleman* and *Hampton,* I think it will be plainly seen that little stallions, with mares suited to them, do get good stock, and much better than most large horses get. For one thing, little mares generally breed larger stock than themselves; so the produce of a small sire is often bigger than its progenitor.

A few failures of large horses may be enumerated in addition to those I have already mentioned. I place first on the list *Prince Charlie*, perhaps the largest, and certainly, I think, the speediest horse of his day. This horse, now defunct, did not prove himself a great stallion; and though he was not without other defects, and serious ones, I attribute the failure in some measure to his immense size. *Wild Dayrell*, *Plenipotentiary*, and *Bran* are three other instances of large horses who, whether in blood or performance, could scarcely be eclipsed, but who proved failures in their progeny. Indeed, the only instance that I can call to mind within the last forty years of a thoroughly good stallion above or about 16 hands high, was *Stockwell*. His brother *Rataplan*, and his half-brother *King Tom*, got many winners; but neither was anything like as good, either at the stud or on the turf, as he himself was, while *King Tom* was the largest of the three, and perhaps was the worst in every respect. It should, however, be remembered that *Stockwell*, though so high, was by no means a leggy horse, but one of the most powerful animals then in existence, or that I ever saw as a race-horse— standing on remarkably short legs. Indeed, it used to be said of him when he stood at the Hooton Paddocks in his later years, that he had the bulk of a brewer's dray-horse; and, apart from the distinctive character of blood-horses in head, neck, and limbs, much the appearance of one. *Stockwell* was, in fact, a sire unique in his way; and the solitary instance of success under these conditions does not condone the failure at the stud, comparative or absolute, of other large horses, as in the case of *Pyrrhus the First, Coronation,* and *Harkaway,* all very large horses. Then we have *Van Tromp* and *Vanderdecken,* out of *Barbelle,* the dam of *Flying Dutchman,* who, although much the smallest of the three, was the best stallion. Again, we have *Abergeldie,* who was 17½ hands high, and a failure at the stud.

The comparison exhibited in the above illustrations of the success respectively of small and large stallions at the stud must, I think, be allowed to be very much in favour of the former. Indeed, I have no hesitation in recommending them as being, in point of size, fit to mate with such mares as I have described and recommended in the previous chapter. This is, however, taking into account only one quality, and that is size. Shape is another essential; and as in my opinion there can only be one good shape, though there may be many bad or indifferent, my ideal may be described in a few words. First, then, it is most desirable that the stallion should stand well on his feet and legs, having plenty of substance. The knees should be good, and

13

the hocks free from curb or any other disease, with good feet, neither too large nor too small. His back and hind-quarters should be good, but if a little shallow in his fore-ribs, like *Venison*, I should not much object to it, if he was not flat-sided but round. His shoulders may be strong and thick at the top, as in the *Touchstone* blood, if only he has action. Otherwise I should prefer him with long shoulders, sloping backwards, with high withers. His head should be intelligent, the forehead broad, and the nose small, with a short neck. This is how I should choose a stallion made to suit rather small mares.

I now come to deal with one very important matter to all breeders of horses; and that is the effect of using an inferior stallion for breeding purposes. It is my conviction that no mare should, under any circumstances, be put to an inferior stallion, for without going so far as to say that many of her foals in after life will resemble the sire of her first produce, and partake of his qualities, good, bad or indifferent, there is no doubt whatever that the subsequent offspring of a brood-mare is and may be affected by the fact of intercourse with some previous sire; and probably this shows the more marked result of the influence upon her produce of the first stallion put to the mare—although I admit this is but conjecture on my part.

What is beyond question is that, in breeding,

some of the most curious, and comparatively unaccountable, facts are seen. I have already mentioned the extraordinary circumstance of the produce of the eggs of the game-fowl resembling the humbler barn-door species, if hatched under a hen of the ordinary kind. This I should not myself be inclined to believe, had I not got it on unimpeachable authority; and though age has made me, perhaps, sceptical as to the truth of every story one is told, it has also taught me to believe in the wisdom of Shakespeare's teaching, that 'there are more things in heaven and earth than are dreamed of in our philosophy,' and therefore not to discard as impossible anything which I learn on good authority, simply because it does not conform with my own previously formed ideas.

One of these curious facts is that I am now dealing with: viz, that the intercourse of a mare with a stallion of an inferior kind may, and indeed does, affect the quality of her offspring got later by another and better sire, so much so that it is said such intercourse of necessity deteriorates all her subsequent foals. The evidence is all one way, and may be regarded as conclusive. One authority —'Cecil'—says on it: 'It is curious to remark that when a thoroughbred mare has once had foals to common horses, no subsequent foals which she may have had by thoroughbred horses have ever evinced any pretensions to racing qualities.

There may be an exception; but I believe I am correct in stating that there is not. It is laid down as a principle, "That when a pure animal, of any breed, has once been pregnant to one of a different breed, she is herself a cross ever after, the purity of her blood having been lost in consequence of this connection."'

To those who may be inclined to regard this as incredible, I would simply say that it is confirmed by actual known facts; that, in short, it may be received as true on unimpeachable authority. It is one of those enigmas far beyond individual opinion or theory, which are hidden from our understanding by the impenetrable veil which often enshrouds the mysteries of Nature. As to the special conditions enumerated by Cecil, I find that, since Cecil's time, there has been a case in which a thoroughbred mare, after being covered by a half-bred horse, produced a winner. This was the case with the Duke of Beaufort's *The Roe*, when she bred *Horseshoe* and *The Cob*. But I never heard of another instance. *Mouche* was put to a pony, but I never heard of the produce winning, nor that any of her stock had subsequently done anything whatever. Other valuable testimony is, however, available on the point, if we are enlightened by it only as to the effect and not as to the cause. 'Sir Gore Ouseley, when in India, purchased an Arab mare, which during several seasons would not breed, and, in consequence, a cross with a zebra was resorted to. She produced an animal striped like its male parent. The first object being accomplished, that of causing her to breed, a thoroughbred horse was selected, but the produce was striped. The following year another horse was chosen, yet the stripes, although less distinct, appeared on the foal. Again Mr. Blaine relates that a chestnut mare also gave birth to a foal by a quagga, that the mare was afterwards put to an Arab horse, but that the progeny exhibited a very striking resemblance to the quagga. The paintings of the animals bred by Sir Gore Ouseley, as also the skins, are to be seen at the museum of the College of Surgeons, in Lincoln's Inn Fields.'

On the same subject Mr. Darwin writes, in his 'Origin of Species': 'In Lord Morton's famous hybrid from a chestnut mare and a male quagga, the hybrid and the pure offspring subsequently produced from the mare by a black Arabian sire were much more plainly barred across the legs than is even the pure quagga;' and he further corroborates what other writers assert on the subject by saying: 'But when a breed has been crossed only once by some other breed, the offspring occasionally shows a tendency to revert in character to the foreign breed for many generations—some say for a dozen or even a score of generations.'

After such testimony, no one, I should think, would contemplate running the almost certain risk of the deterioration of a valuable mare by putting her to an inferior stallion.

In a subsequent chapter on 'The Thoroughbred Stallion' I shall have something to add on the question of age and other matters.

CHAPTER XV.

THE THOROUGHBRED * MARE.

Respective influence of sire and dam on their offspring; discussion—Instances—Suitability essential.

Opinions as to respective ages of sire and dam—Some notable first foals—Youth evidently no drawback on either side.

Inequalities of size considered—Large animals have poor progeny—Remedying of defects by suitable selection—Different methods employed by breeders—Opinion of an expert—Experiment of the late Earl of Derby—Size should be improved by gradual selection.

Curious fact that mares breed best from particular stallions—Cases in point—Breeding a winner proves the suitability of the cross—*Defenceless, Flying Duchess,* and *Elcho's* dam—Innumerable proofs of my contention.

The question whether young mares are better than those which have raced long, or the reverse—My view that we must always look to pedigree—Three examples on each side—Solution of the problem.

BOUNTIFUL Nature is all-powerful, and her laws cannot be transgressed with impunity; we may

* There may be readers of this book who are ignorant of what is technically meant by the term 'thoroughbred,' which is very frequently misapplied. I may define it for their benefit in this way: No horse or mare can claim to be 'thoroughbred' unless

assist, and we may even bring about variety in her productions; but more than this it passes human ability to accomplish. Injudicious interference with the due course of Nature's processes is more likely to spoil and maim her operations, which, if left alone, might have resulted excellently, than to improve them by running counter to her immutable laws. Hence, in whatever direction we essay to remedy defects of parents in their offspring, we must be satisfied to do so by slow gradations. The full benefit we seek to attain, and may expect to achieve ultimately, is not likely to be gained at once.

The influence that sire and dam respectively have upon their progeny is a matter of much importance. It is possible to form an opinion on the subject, but not possible to prove it. If anything like a formula could be laid down, the exposition of which were practical and credible, touching the exact relations of either parent to the offspring, then we should have a most desirable clue to guide us in breed-

---

his or her name is entered in 'The Stud-Book,' which constitutes a sort of patent of nobility. The entry cannot be made unless both the animal's parents are also in 'The Stud-Book,' and theirs again, and so back for a hundred years. Any thoroughbred's ancestry can thus be clearly traced up to a few original progenitors. A thoroughbred, therefore, must have been got by a thoroughbred stallion out of a thoroughbred mare, and all their names must be entered in 'The Stud-Book,' showing the entire pedigree. No other horse can legitimately claim the title.

ing good horses. However, I fear that no certain principle can ever be propounded by which we could determine beforehand whether the qualities and characteristics of the sire, or of the dam, should most preponderate in their offspring.

I used to think that the mare would exercise the greater influence over her foal, both as to its physical formation and as to its abilities as a racer, and in every other respect. If this were so, then a mare that could breed a foal as good as herself to one stallion, should do so to another, however differently he might be bred. If, on the contrary, the characteristics of the sire generally predominated in the foal, then he should get good horses out of any mare.

Now, *Penelope* bred her best horses to one stallion, *Waxy*, yet we find many mares in 'The Stud-Book' which did not; for instance, *Phryne*, *Pocahontas*, *Barbell*, *Alice Hawthorn*, *Monimia*, and *Ellermire*. It is noticeable that the stock got by some of the most successful stallions out of a variety of mares differs in shape, size, and breeding, although affording good runners. *Touchstone*, *Newminster*, and *Orlando* got many winners out of mares which never bred any before; other instances of the same thing are found in the cases of *Stockwell* and *Rataplan*, and later, of *Hermit*, *Galopin*, and *Hampton*. Therefore we have evidence that good mares can breed good horses, and that good stallions can get the same from any mare, provided it be thoroughbred.

From this I would not have breeders believe that they possess mares equally well suited to all stallions alike; or that they have stallions as well suited to one mare as another. That is by no means my contention; for it is but reasonable to opine that well-bred mares, put to stallions that do not suit them, will not breed such good horses as if put to stallions that do suit them, in regard to shape, ability, and all desirable qualities.

A theory that many give weight to is, that either the horse should be older than the mare, or the mare than the horse. Equality of age, according to this, is bad. But there is no consensus of opinion among those who hold the theory as to which of the two, horse or mare, should be the senior, so as to ensure good stock. Some incline one way, some the other, both sides bringing numberless examples forward to prove each its own case.

A mare is said to breed weakly stock if she be either very young or very old; and a stallion is thought to get his best when he is tolerably old. These are but speculations, however, and need much stronger confirmatory evidence than has yet been brought forward in support of them.

We are told that all the following were first foals: *Mark Antony, Conductor, Shuttle, Pope,*

Filho da Puta, Pericles, Doctor Syntax, Ugly Buck, Pantaloon, Melbourne, Lath, Miner, and *Wild Dayrell;* as also *Touchstone* and his sire, *Sultan.* After so many noted examples, surely the firstborn, even of young mares, are not to be despised ? I can see no reason myself why a young mare's first foal should not turn out as big and good as anything she may afterwards have by the same stallion.

We have also quite as many examples to show that a stallion, in the first and second seasons he is at the stud after racing, can get as good a foal as he may later in life. Indeed, some of our very best horses were got in their sires' first season of serving.

Rams of seven or eight months old, before they are full-grown, consequently, get the best and largest lambs. They are, therefore, put to old ewes, while old rams are put to the young ewes ; by which means the risks of lambing are lessened. Such is the system adopted, and the reason given for it, by many breeders of sheep.

Large stallions and small mares, or *vice versâ*, seldom breed anything as good as themselves ; all matching of extremes is generally bad in result, and therefore should not be resorted to. 'For racing purposes,' says Cecil, 'I cannot advocate over-sized horses. Immensely large ones have, on all occasions, disappointed the hopes of their owners ; they have neither realized great superiority on the turf nor in the stud. It is evidently much more difficult to obtain true symmetry in over-sized than in moderately-sized animals. Very large mares are of all others the most likely to occasion disappointment. On some occasions they will produce foals equally large or larger than themselves. In other years they will give birth to such as are under-sized, and in most cases their stock is puny and weak.'

Except that large race-horses are sometimes good on the turf, I am quite in agreement with the above-quoted opinion. I do not remember a single example of a very lengthy mare producing a good foal by a very short, compact horse. The two extremes are too great, though I dare say there may be exceptional cases. I do not believe in the goodness of very large stallions—but of them more anon—nor do I like large mares. No doubt, however, there may have been instances of such animals having been successfully mated with horses somewhat smaller than themselves, and coming from a small stock. Such an instance we have in *Stockwell*, whose sire was *The Baron*, rather a small horse, according to the accounts we have of him, as were many of his descendants. *Pocahontas* was said by a sporting authority to have been a small mare, but my recollection of her when she ran for the Oaks is entirely different. I believe she was 16 hands high, and that she was a loose-made sort of animal; hence she would be better suited by *The Baron* than by a bigger horse like *Harkaway*; for, as it proved, *King Tom* turned out worse than either *Stockwell* or *Rataplan*, both as a racer and at the stud.

Technical science teaches us that we may, from the same sire or dam, breed a race of giants or dwarfs, in a few years, by means of suitable and judicious crossing. We may occasionally correct or modify defects of shape in the mare by putting her to horses chosen in such a way that we may calculate the sire's conformation will remedy faults of the dam's in the offspring of both. In like manner mares may be selected to correct faults in the stallion. But, of course, whenever practicable, it is far better to breed from animals whose shape needs no correcting, or, at least, not enough to suggest the selection of mates for them on account of such especial requirements. If, for instance, a mare with good shoulders be put to a stallion similarly well endowed, the probability that their offspring will have good shoulders too is infinitely more likely than it would be if but one of the parents possessed that particular desideratum. Precisely the same principle holds good in respect to any other feature of the body, and, in fact, applies directly or indirectly, and in every conceivable way, to the whole economy of the animal.

A few illustrations of the success or failure of the

different methods generally pursued by breeders may be advantageously alluded to. As in training, so in all matters connected with breeding; each individual advocates some pet theory of his own in a plausible manner 'which, seasoned by a gracious voice, obscures the show' of evil;' and, however erroneous the hypothesis may be, its folly only practical knowledge can detect and fearlessness expose. Here is the opinion of Professor Pritchard, an admirable expert in whom I have great confidence, and whose views I would recommend to the careful consideration of every breeder. 'To put a soundly-constituted horse to a roarer, or a horse with a thorough-pin to one with good hocks, is a huge mistake. For, though the disease may lie dormant in the offspring for a time, any exciting cause will be sure to develop it, rendering the animal useless.' As to the soundness of that doctrine there can be no room for doubt.

The late Earl of Derby, a nobleman of vast experience and great knowledge in breeding, put *Canezou*, a very large and good mare, to a small stallion, in order to obtain a moderately-sized foal. What he did get was a great overgrown brute of 17 hands high, which was never put into training, I believe. I saw the animal when he was put up to auction at Doncaster, being then two years old and unbroken; but I never saw or heard of him afterwards. The case helps to show that all extremes are bad. I hardly ever remember seeing a very good race-horse the produce of a very large sire out of a dam under the average size, or one by a little horse out of an overgrown, unwieldy mare. Yet I admit there may be exceptions to this rule, as to all others. *Monstrosity* was one such. As her name implies, she was a great overgrown animal; she was sent to the stud as a two-year-old and put to *Venison*, a little horse, by whom she had the well-known *Ugly Buck*, winner of the Two Thousand. That he was a very exceptional case is unquestionable, for *Ugly Doe*, own sister to him, was a complete failure, and *Monstrosity's* success ended with him; which may be taken as showing that the rule holds good, in spite of occasional exception to it.

If a brood-mare be under-sized, she should be put to a stallion somewhat, but not too greatly, larger than herself, and the same system carried out with her offspring; so, in the course of two or three generations, we shall be most likely to attain the object of our desires—an increase in size without deterioration in quality. On the same principle, if the mare be over-sized, we should select a rather smaller stallion, so that the same effect may gradually be brought about, and the produce in either case prove a triumph of genius.

A curious fact, which must be familiar to most breeders, is that a mare will often breed better horses from some one particular stallion than from any other. Yet, where such has been observed to be the case, efforts are seldom made to preserve the connection which has proved so favourable. Breeders will go on trying different stallions, which, however well they may seem to be suited to the mare, cannot get good horses out of her. Thus half the life of many a good mare is thrown away in consequence of these useless experiments, as I call them, on account of the ignorance or thoughtlessness of an inexperienced or self-opinionated breeder, by continuing to put her to other stallions than the one which has been found to suit her, and which is very likely the only one in existence that does.

A case or two in point may be serviceable, as tending to show the truth of the last assumption.

Leda was bred in 1859; she was put to three different stallions, and had five foals by them, none of which were good for anything. In 1869 she was put to *Young Trumpeter*, and by him had *Bugler*, who, as a two-year-old, proved a really good horse, if not the best of his year. After this I sold her to a celebrated and successful breeder, who put her to many of the most fashionable stallions of the day year after year; yet she bred nothing by any of them but what was almost worthless. Now, it is probable that if she had been put to *Young Trumpeter* again, time after time, she might have bred several good horses.

Should a mare breed a winner, if only of a small race, at or about the commencement of her breeding career, it gives presage of better things; and the same cross should be persevered with, even though a bad horse or two may now and then result from it. *Defenceless* bred *Haunch of Venison* by *Venison*, a winner of one or two small races; and afterwards she was put to *Bay Middleton* and to *Kingston*, and other fashionable stallions for years in succession. The last year she was put to *Kingston* she had *Caractacus*, who won the Derby in 1286. She had nineteen foals by eight stallions; but none of the latter had *Venison* blood in them, except *King John*, son of *Kingston*, and he was not thought likely to get a race-horse by anyone but his infatuated and persevering owner. Out of the nineteen foals *Defenceless* bred, only the two above-named were good for anything—the one by *Venison*, and the other by his son *Kingston*.

A similar example is shown in the case of *Flying Duchess*. She was put to *Vedette*, and bred *Pevensey*, a winner, afterwards *Vex*, and then *Galopin*, to the same horse; but her produce by five other stallions was bad.

Elcho's dam's first foal was *Amati*, who was by *Womersley*, and had been a winner. Afterwards she bred *Elcho* and *Bosh*, his brother, by *Rifleman*. This mare (*Elcho's* dam) was unnamed, I believe—at least, I do not know if she had any name. She was bred in 1847 by *Sleight of Hand* out of *Hamptonia* by *Hampton*. She had thirteen foals by eight other stallions; yet they were all very moderate. The best of them was probably *Angus*, and he was by *Newminster*. Then *Touchstone* (a first foal) and *Satirist* were own brothers. The same relationship existed between *Promised Land*, *Happy Land*, and *Jerry Kent;* also between *Rubens*, *Selim*, and *Castral*, and between *Stockwell* and *Rataplan;* whilst, before breeding *Voltigeur*, his dam, *Martha Lynn*, produced *Barnton* to the same sire. *Whalebone* and *Whisker* were also brothers.

Innumerable instances of the same kind could be adduced; but I must refrain from overdoing the matter, for I fear that, like Gratiano's reasons, they would not be worth the search when found and produced, though they might be, like Sancho Panza's proverbs, 'more numerous than those of the Greek commentators, and equally admirable for their sententious brevity.'

Mares that have proved bad racers over any distance, but which come from a long line of good horses, may still be safely bred from. Such mares, indeed, often do better at the stud than do others which have excelled them on the turf, and possess equally good pedigrees. This is a fact that cannot be accounted for. In the one case, some people think a mare's powers are exhausted by long racing; so that, if she goes to the stud at a later period of her life, she is less fitted for breeding. Others hold that a mare which proves useless as a racer, or does not race at all, and is early put to the stud, has all her energies conserved and devoted to the one purpose of breeding. In the face of such conflicting ideas upon a subject otherwise unaccounted for, it is necessary to move with caution, and not to form an opinion without duly weighing all that may be advanced in contravention thereof.

I consider it absolutely essential that, whatever may be the good qualities of a mare we select, those qualities should have been possessed by her parents and 'forebears' in greater or less degree. Here are some examples which illustrate both views set forth in the preceding paragraph.

Paradigm ran but twice, and never after she was two years old. She was put to *Kingston* then —in 1865—and produced *King-at-Arms*. Subsequently she was the dam of *Lord Lyon*, *Achievement*, and of other good horses, proving herself in every way one of the best of brood-mares.

Miss Twickenham ran but once, and then was put to *Orlando*, by whom she had *Teddington*.

Maid of Palmyra never ran, but was the dam of some very good horses.

On the other hand, we have *Beeswing*, who ran over most severe courses, seldom under two miles, for six years. She did not retire from the turf till she was nine, after winning forty races. She

14—2

bred her first foal when eleven, and afterwards had seven more, among them *Newminster*, winner of the Derby, and *Nunnykirk*, his brother; she died at the age of twenty.

Few mares have been harder worked than *Alice Hawthorn*. She ran in no less than seventy races, winning forty-two of them. Yet she bred *Thormanby*, a Derby winner; *Oulston*, who was sold by that best of salesmen, Mr. Padwick, for £8,000, and eight others. She was put to the horse when eight years old, and died at twenty-three.

Crucifi , also, could not be said to have had an easy time of it. She ran eleven races in two years; yet she bred *Surplice*, winner of the Derby and St. Leger, *Cowl*, and other good horses.

Here is proof for both views above-stated, and equally proof against both. Evidently we must look further for the reason why one mare breeds good stock and another bad. Running long races, or remaining long on the turf, it is here seen, do not necessarily prove prejudicial to a mare's reproductive powers. Equally, being put to the stud early, sometimes without racing at all, may be attended with every desired success. To my thinking, the solution of the problem will be found by referring to the ancestry of the respective mares. In that, and in properly mating them, I am convinced, will be discovered the true explanation of this interesting question.

CHAPTER XVI.

THE THOROUGHBRED STALLION.

Former partiality for the strains of *Highflyer, Eclipse,* and *Matchem*—Present liking for those of *Touchstone, Voltigeur,* and *Stockwell*—The latter the best horses yet seen—Their pedigree and descendants.

Evils of breeding from your own stallion; exception and proof—Rank fallacy of the practice demonstrated—The experience of various gentlemen—Numerous instances—The unavoidable deduction therefrom.

Age of the stallion—Capriciousness of evidence—No rule can be fixed upon—Various opinions.

The Queen's Plates—Prohibition of geldings—I demur to this—Many inferior stallions should be cut—Encouragement to geldings might improve our breed—Mr. Craven's opinion—The matter gravely considered—The knife necessary.

Anomalies in breeding—Trainers' systems—Public running a farce—A typical trainer—His failure at long distances—Speed *versus* stamina—The historian's duty.

FORMERLY, the best judges of breeding looked upon strains derived from *Highflyer, Eclipse* and *Matchem* as the best to follow. Now, and for a century to come, it is probable we shall have to regard in the same way the blood of *Touchstone, Voltigeur,* and

Stockwell. From them the best of our present horses are descended; and if we can trace back our blood to those sires, it will be sufficient without going beyond that date.

No doubt we have a good game breed coming from *Dick Andrews* through his son *Tramp*, and from him through *Lottery, Sheet Anchor, Weatherbit, Beadsman, Rosicrucian;* and we have a softer strain also from *Sir Peter, Walton, Partisan, Gladiator, Sweetmeat, Parmesan,* and *Cremorne*. Nevertheless, the three best stallions this generation has seen—perhaps, indeed, the best *ever* yet seen—are *Touchstone, Voltigeur,* and *Stockwell*. They are, and always will be, looked upon, and justly, with as much veneration as the *Godolphin Arabian, Eclipse,* and a few more, were by our forefathers.

I had intended to present a table showing the pedigrees of our thoroughbreds from the Eastern horses, Arabians, Barbs, and Turks, from the time of their introduction before, in, and since 1750. However, in connection with what I have just said, there would seem to be more use and interest in showing how *Touchstone, Voltigeur* and *Stockwell* are related by their common descent from *Eclipse,* and how the line continues from each of them. I therefore subjoin a table of pedigrees, derived from the invaluable 'Stud-Book' of Messrs. Weatherby (vol. xv., pp. 714-723), referring the inquiring reader to the same for further information.

ECLIPSE (foaled 1764)—

Sire of	Sire of	Sire of
Pot-8-os, 1773—	King Fergus, 1775—	Pot-8-os, 1773—
Waxy, 1790—	Hambletonian, 1792—	Waxy, 1790—
Whalebone, 1807—	Whitelock, 1803—	Whalebone, 1807—
Camel, 1822—	Blacklock, 1814—	Sir Hercules, 1826—
TOUCHSTONE, 1831—	Voltaire, 1826—	Birdcatcher, 1833—
Newminster, 1848—	VOLTIGEUR, 1847—	The Baron, 1842—
Hermit, 1864—	Vedette, 1851—	STOCKWELL, 1849—
St. Blaise, 1880.	Galopin, 1872—	Doncaster, 1870—
	St. Simon, 1881.	Bend Or, 1877—
		Ormonde, 1883.

In order to breed good horses, a man should not keep his own stallion; at least, not in general. Mr. Chaplin has done so, and still continues the practice; but in his case the stallions kept are two Derby winners, exceptionally good horses. Though the plan has succeeded with him, I think it is not a wise one to follow generally. It failed with Lord George Bentinck, who kept *Bay Middleton* and other good stallions, and ruined the best stud of mares then in existence. Much the same might be said with regard to the Rawcliffe stud and the Middle Park stud; and many small breeders have failed to succeed for the same reason.

For, to start a stud with one or two good racehorses and fifty or sixty mares to put to these, their own horses, is a rank fallacy, and has been proved so over and over again in too many cases, both of private studs and of others belonging to public companies. Rawcliffe was a great failure, despite of having *Newminster* and other good stallions, management on such principles proving no good. Cobham possessed *Blair Athol*, and it was the same thing there, as likewise at Fairfield and Middle Park.

It is true that Mr. Blenkiron bred two Derby winners. One of these, however, *Caractacus*, was only a moderate horse at the best, compared with other winners of the great race. The other was a good one, but was not got by Mr. Blenkiron's own stallion. In this case he had done what I have already recommended all breeders to do—sent the dam to another stallion. He had selected *Newminster*, then at the Rawcliffe Stud Paddocks, York, a horse that could get a good foal anywhere except in his own harem, the mares in which were not suited to him. *Hermit*, taking him both as a racer and a stallion, was the best horse Mr. Blenkiron had. It was the same in the case of *Bay Middleton*, who was able to get *Andover*, *Hermit* (foaled in 1851), *Flying Dutchman*, and *Mary Copp*, for other people, but failed to do as well by any of the half-hundred best mares in his owner's stud.

We had a similar example of this in *Vedette*. He had a large number of mares out of the Diss stud, during the years he stood there, but none of his stock out of them won classic races. I know he got *Galopin*, who won the Derby, having been bought as a yearling by Prince Batthyany for five hundred guineas at the Middle Park sale in 1873. But when Mr. Simpson's stud was broken up, Mr. Taylor Sharp bought *Flying Duchess* and her foal, afterwards re-selling them—as appears in the 'Stud-Book'—to Mr. Blenkiron. Therefore it is plain that the honour of breeding *Galopin* does not appertain to the Middle Park stud. So far, then, my contention remains not disproved—namely, that few good horses are got by keeping a stallion especially for a large stud of mares to breed from exclusively.

Sir Joseph Hawley's *Charleston* was a failure, although it is true his *Weatherbit* had success. My own stud was a failure, while I followed the same plan. I sent one mare to *Flatcatcher*, and bred *Catch-'em-Alive*, and another to *St. Albans*, and she bred good fair horses. My mares were good ones, and my stallions should have suited them, but they failed. Mr. Cookson, with *Buccaneer*, and Sir Tatton Sykes with *Womersley* and *Rifleman*, had exceptional successes; to which may be added the Duke of Westminster's experience. But these are all such, I think. Let us now see some of the results obtained by private breeders who did not keep stallions of their own.

Mr. Wm. Sadler bred *Pretender*, winner of the Derby, and *St. Hubert*. The present Sir Tatton Sykes bred *Frivolity* and *Doncaster*, winner of the

Derby. Mr. Whitworth bred *Mendicant*, winner of the Oaks, and foundress of Sir Joseph Hawley's stud. Mr. Plummer bred *Alice Hawthorn*, and her son *Thormanby*, winner of the Derby. Mr. Stephenson bred *Virago*, and *Voltigeur*, winner of the Derby. Mr. I'Anson bred *Blink Bonny* and *Blair Athol*, both Derby winners.

Here is a list of Derby winners bred by gentlemen who, even if they kept a stallion at all with their own studs, at any rate, in these cases, had sent the respective dams to others: *Coronation, Andover, Wild Dayrell, Pretender, Doncaster, Blink Bonny, Blair Athol, Lord Lyon, Pyrrhus the First, Hermit, Thormanby, Voltigeur, Sir Bevis, St. Blaise, Musjid*. In the same way were bred *Lord Clifden, Rifleman, Geheimniss*, besides a host of others, all good horses, too numerous to mention.

I venture to say that the breeders of these horses did not at that time possess one-third the number of mares that were in the Rawcliffe, Cobham, Fairfield, and Middle Park studs. Yet all these large establishments, perhaps the biggest in England, failed among them to produce more than two Derby winners. Only one of these, *Caractacus*, was got by a stallion belonging to the stud in which he was bred, namely, *Kingston*. On the other hand Mr. Wreford, so long as he sent his mares to different stallions, bred with the greatest success, and for many consecutive years.

When, however, he took to breeding from his own stallion, *Sultan Junior*, his luck left him, and ruin succeeded. From all this, therefore, I think it is plainly obvious that to habitually breed from one's own stallions is an undoubted mistake.

It is an undisputed fact that many of our best stallions have got as good stock in the first year they were at the stud as they ever did subsequently. The first year that he covered, *Touchstone* got *Blue Bonnet*, winner of the St. Leger, besides many other winners. He got *Soapstone* when he was twenty-seven. Similarly, *Pyrrhus the First* got *Virago* in his youth, but nothing so good afterwards.

On the other hand, the sire of *Eclipse* was in his fifteenth year when he got him, the dam being the same age. The *Godolphin Arabian* got nothing good after he was sixteen. *Sir Hercules*, however, died at the age of thirty, and the three previous years had got *Gunboat*, *Lifeboat*, and his brother, all as good horses as he was himself.

The facts relating to age are curious, and are not always easy to interpret as a guide. Lambs invariably get bigger and better stock than mature rams do, as I have already said. Boars and bulls also, before they are much more than half grown, will get stock not lacking in size or stamina.

Mares are not less precocious than horses, and there are numerous instances on record of their breeding, in the first year after being covered, stock as good as any they might afterwards have produced.

I am of opinion that, as long as a stallion remains vigorous and capable of procreation, he is as likely to get good horses as not; that, even in old age, his stock may be as good as any he got in his youth or in his prime.

No rule can be laid down with any degree of certainty as to whether it is best for the mare to be younger than the horse or to be older. In either case the mare may foal her best stock, and in either the horse may get his. There is nothing to direct us on this point. Instances there are in plenty, some of which I have referred to, proving that horses both can and do get good stock to the last, and may not get their best in youth. It is rare, however, when they do so after the age of twenty.

The age at which a stallion is best fitted to commence at the stud is a moot point. There is much difference among breeders with regard to it. Some think that a three-year-old stallion will get as good stock as he will at any time up to the age of twenty. Others hold that no horse can be fit for the stud until he is seven, or a six-year-old at the earliest; and many affirm that after twenty his stock will be valueless. I see no reason why a three-year-old—who would, under racing rules, be actually somewhat older, probably—should not get as good stock then as ever he would afterwards. *Wild Dayrell* served his first season when he was four years old. The reason why our stallions are usually not put to the stud until they are older, is simply because it is thought more profitable to keep them racing. It is only those that break down, or which leave the turf early for some other sufficient reason, that are put to the stud while still young.

From time to time, various suggestions have been submitted to the consideration of the Master of the Horse with regard to making the Queen's Plates more attractive to good horses, and thus improving the breed. But such attempts have failed, for twenty or thirty years past. In view of the large stakes that can be run for elsewhere, almost any day of the week, it is not easy to see how the Queen's Plates are to be taken more into favour in the future than in the past. The latest alteration has been the prohibition of geldings from running for them.

Now, this regulation was, I venture to say, by no means a good one. It was, I think, unwise to make it impossible for a gelding to win any such prize. I would far rather have heard it decreed that these races should be run for by thoroughbred geldings only, which had been castrated as yearlings. We should then have really witnessed some improvement in our breed, and that rapidly. For hundreds of brutes would be cut, on the chance of some of them winning a Queen's Plate now and then; and thus the country would be rid of a class of stallions now everywhere propagating stock more useless than they are themselves.

More races ought to be instituted for geldings to run in, whether the prizes be provided by the Jockey Club or otherwise. The creation of them would quickly call into being committees and clerks of the course to provide for the especial management of them. I mention this to show that the Queen's Plates might be given so as to benefit and improve the race of thoroughbreds in more ways than one, and in better ways than the latest suggestions—to give prizes for stallions at agricultural and other shows.*

On this point, Mr. W. G. Craven says: 'The Jockey Club should be called upon, after three years commencing of such a stud, to give stakes to be run for at four years old by geldings; in fact, to give some temptation to owners to alter many horses, which, being too bad to win races, are now poisoning the country by travelling in the spring.' Here is the gist of the whole matter—geld more freely,

* Since this was written the Queen's Plates have been withdrawn, and are now to be given as prizes at the Agricultural Show. If half the value were given to geldings, on the principle I have suggested, the fact would certainly lead to improvement in the qualities of our stallions, as I have above set forth.

and we should have fewer and better stallions ; and, as a natural consequence, a superior breed in a few years. This plan I have strongly recommended for adoption in thoroughbred horses, and it is as applicable and as needful to all other descriptions. We also hear from other quarters that there is in the country a scarcity of horses and mares fitted for breeding horses of a useful description ; and that, in reality, we are as much in want of horses for home defence as for foreign service.

Now, if it be conceded that we are overrun with indifferent horses—horses withdrawn from the turf simply because they are useless there, and of no account anywhere else, either, if we would only recognise the fact—how, I may ask, are breeders to select from among such animals stallions likely to suit their mares ? Since public performance is the chief, though not the only, criterion, and as these have only an indifferent record in that respect, if any at all, how is anything like proper selection possible ? Is it not obvious that the only remedy is the free use of the castrating knife ?

We often see supposed non-stayers get good game horses, much to everyone's astonishment ; just as we also sometimes find a slow mare the dam of a speedy animal. One might account for the first by supposing that the ideal short runner is, in fact, a good, honest, soundly constituted horse, able to stay any distance, but who has been placed at as much disadvantage at the stud by not having mares to suit him, as he was on the turf by being run out of his course. A precisely similar hypothesis may be formed in the case of the slow mare I have instanced. Here, then, are difficulties which, if superficially looked at, would seem to made breeding as happy-go-lucky an affair as if we make our selections of mares and stallions blindfolded. That it is not so, or, at any rate, need not necessarily be so. I think I have abundantly and conclusively demonstrated.

There is another thing. All trainers do not follow the same system with their horses. Now, it would seem apparent that only one way can be the right way, and that, therefore, all other ways must be wrong ways. Hence follows, doubtless, the very different performances of a horse when in the hands of different trainers. Much the same might be advanced with regard to jockeys ; and, in saying this, I intend to pass no individual reflections either on them or on trainers. But the fact remains that, in consequence of the capriciousness of methods employed, public running is little better than a farce, so far as it may be held to demonstrate the relative capabilities of horses. For they, like men, often 'gain a reputation without merit, and lose it without deserving.'

Perhaps an example from my own knowledge may serve to illustrate what I have just advanced.

I knew a trainer who had for some thirty years a large stable of horses, sometimes including more than a hundred. Many of them were of the most fashionable and expensive breeding. I cannot call to mind that he ever won a single race of two miles' distance, or over that; although he often ran horses over such courses. He might have done so, perhaps; but my firm impression is, that he never trained the winner of a race of that distance. He won very many races of five or six furlongs— his favourite distance—though mostly small ones. Occasionally he may have achieved a mile race, or sometimes one of a mile and a half; but if he ever did get the last, they were very few indeed.

This was an accomplished trainer and most estimable man in every way, generally considered to stand at the very top of his profession—if I may be allowed so to term it. Suppose we grant that his predilection for buying descendants of speedy ancestors often got the better of his judgment, as might be the case with any fallible mortal; yet he may have purchased, and probably often did, many a real stayer without knowing it. I dare say he had in his possession, at one time or another, many such animals. For there must have passed through his hands in all perhaps a thousand yearlings, and as many more older horses, some his own property, and some those of his numerous employers. Among the latter there would doubtless be some who preferred a stayer to a speedy jade, and who would rather have won a long race over a Cup course than a short one. Yet the facts were as I have stated, and would seem to throw suspicion of faultiness upon his methods of training. On the other hand, there are certainly trainers who seldom or never win any race of consequence over a short course, yet who achieve distinction by their well-earned victories over a distance of ground. And it is by due reflection on these facts that we are placed in a position to deduce some explanation of the deceptive running so commonly seen.

No doubt it was pleasanter for Livy to dilate on the grandeur, wealth, and power of Rome than it was for Gibbon to chronicle its 'Decline and Fall.' But however disagreeable it may be for an historian to detect and expose error, it is none the less a necessary task, and must be executed with impartial candour. At the same time we must be careful not to enlarge or dwarf matters, whether they be favourable or the reverse. For to accord undue praise to actions worthy of admiration is as unseemly and unjust as to magnify foibles into gross faults. Either extreme ought to be carefully avoided by one whose only aim is to faithfully portray things according as he finds or has knowledge of them.

CHAPTER XVII.

MATING.

Meeting of the sexes—Time to put mares to the horse—The mare should be hobbled when covered; reasons.

The moment of conception no concern of the breeder's—Repetitions of service to be avoided—Mr. Sadler's experience in this respect—A case that occurred to me.

Virgin mares to be cautiously dealt with—Attention necessary—Bringing them to the stallion—Care to be taken not to exhaust the horse's powers.

Thoroughbred mares—Consideration of time of mating with regard to racing rules—The fourteenth of February—Period of gestation of mares—A popular theory—Service of barren mares—Likelihood of procuring pregnancy—Modes of causing the mare to be stinted.

Two systems of covering—My view—Instance in the case of bitches—An expert's opinion.

Treatment of mare after service—Turning—Frequency of covering; when desirable and not—Indications of being at use—Opportunity to be seized—Disadvantages of late-born horses—A year's rest for brood-mares; my opinion on that head.

IN previous chapters I have described at length the shape, size, constitution, and other essentials which should be possessed by stallion and mare, whereby they are best fitted for procreation, and for the propagation of a race more numerous than themselves, and one more amply endowed with the most approved qualities. I have said all that it is needful to say on this score, and I now come to the actual meeting of the sexes, and the effects of their connection.

The best time for mares with foals to be again covered is nine or ten days after the birth of the latter, or within a day or two of that time. Usually they are then in a fit condition for receiving the horse, and probably may do so under more favourable auspices than at any other period. The tenth day after foaling is preferred by many breeders, but in this we must allow circumstances to guide us to some extent. The proper moment can only be determined by a wise and cautious consideration of the state of each individual mare. In the case of a thoroughbred mare, too, consideration must be had of the time of year, and whether it be not necessary on that account to withhold her from the horse until another pride.

Here let me just say that no mare should be covered without first being hobbled. For, however easy and apparently quiet a mare may seem when at use, there is always the risk of her kicking at the horse after having been served. I once saw a curious case of this. *Merry May*, by *Orlando*, was being covered. Just as the horse was leaving her she kicked out at him most viciously, and, had she not been hobbled, she would assuredly have done him serious injury, perhaps have ruined him altogether. She was not served again that year, but proved in foal. Such cases teach us to be always very careful in regard to this.

The disputed question as to whether conception follows instantaneously upon the act of coition, or whether the female uterus is not impregnated until some time after the semen of the male has passed into the vagina, is not of any consequence to our present inquiries. Specialists differ on the point, some affirming that the mare is fertilized in the moment of coition, others believing that the said effect does not follow for some time, perhaps days, after the act. However such researches may interest students of physiology, they can be of no use or benefit to the practical breeder. The actual process of Nature in bringing about fertilization through the contact and admixture of certain secretions is, to my thinking, one of those of her mysteries not given to even the most talented of men to discover. We know that a conjunction of the sexes is necessary to the end of producing offspring; and that is about all we do know, or ever are likely to know, I think, and it suffices for the purpose in hand. I shall, therefore, avoid speculation on the subject, confining myself to details really within our knowledge and of true practical utility.

Many mares get covered over and over again, perhaps being in foal all the time, and eventually producing offspring begotten from the first service. This should teach us not to worry them with the dallying of the stallion, unless we see unmistakable signs that they are at use; and even then some circumspection is necessary. Here is a case in point:

Mr. Sadler supposed that one of his mares was not in foal, and his stud-groom was of the same opinion. Everything indicated she was not, and, as she was at use, they both thought there could be no mistake in the matter. She was accordingly put to *Venison*, who covered her. The very next day she gave birth to a foal. It did not live many hours; but whether it died in consequence of the mare having been covered so shortly before she bore it, or from some other cause, I am quite unable to state.

A not dissimilar case once occurred in my own experience. I had a young mare which we thought was barren, and proposed to have served again. However, covering was deferred, and in a few days she foaled, though up to the moment of her doing so there was not the slightest indication of her pregnancy. Nor should this instance be regarded as an altogether exceptional one. Young mares often do not betray the slightest sign of their pregnancy until they are just on the point of foaling, or, at least, until they are very near it.

Mares that have never been covered before must

be 'introduced' to the horse, as it were. They should be shown him several times, and so gradually accustomed to his presence and familiarized with his dallying. Very often they will show no signs of being at use until habituated to the proximity of the stallion. I believe that many never get served at all, simply because due attention is not paid to the first meeting. Some mares are taken to the horse, and on showing vicious inclinations are immediately taken away again; whereas, if left with him a little longer, or brought up to him again and again, their natural desire would be aroused. This happens more frequently with young mares than with old ones, and is, I suppose, chiefly due to their timidity.

It happens sometimes that a mare may be at use one day, and yet have to be put back and covered on the next. This, of course, is owing to the stallion having had his full complement of mares already on the first day; when to use him again would weaken and harm him. Many people think that a mare served on the second day she is at use, instead of on the first, is more likely to be impregnated and have produce; reasons for which belief I have elsewhere noted.

The time when thoroughbred mares should be served must be considered in relation to existing rules of racing in respect of age. The 14th of February is the earliest date I can recommend, and even a little later may be advisable. It will not do to have a Christmas present instead of a New Year's gift; and that risk must assuredly be incurred if we put our mare to the horse too early. If she foals too soon, she must be kept back till on pride within 'measurable distance' of the suitable date I have indicated.

The normal period of gestation for mares is eleven months; but as is the case among all species of mammals, from human beings down to the humblest quadruped, some variation of the regular period is not infrequent. A mare may not carry her foal the full time, but bring it forth a week, a fortnight, or a month earlier. On the other hand, she may carry it longer than eleven months, even for twelve. In this latter case some breeders affirm that she is sure to bear a colt. I have seen no sufficient proof of that fact, if it be one, and I do not attach any weight to the assertion. That by the way. What I would urge is, that we must not lose sight of the more important and actual fact, which is, that a mare covered prior to the date I have mentioned might very possibly be delivered of her offspring before the end of the year, or, at any rate, so early in January as to be unpleasant. Wherefore, so long as the present rules as to age remain in force, I would not put a mare to the horse before the latter half of February.

This being held in view accordingly, at any time after the middle of February barren mares may be brought to the horse and tried. As soon as they are at use they can be served, and this may be done until they are stinted, or thought to be so. They are usually more difficult to stint than foaling mares. This time of year is, I think, quite as favourable for the prospect of getting a barren mare pregnant as any other. Some breeders take a different view, holding that such mares are more likely to be got in foal in April or May, after they have been out at grass a few weeks. There may be something in the contention, but I have seen no reason to place more reliance on the one plan than the other. We may succeed in either case, or we may not; that is about all there is to say with regard to it.

Mares that are difficult to stint after covering have been treated in various ways, so as to procure that result and render them prolific. Bleeding the mare immediately after the horse quits her is one method. Placing a bunch of nettles, furze, or briar under her tail, with the object of so preventing expulsion of the generative fluid, is another. I have seen these practices tried, but without any sufficiently decided result; and I should say that they would be 'more honoured in the breach than the observance.' They belong to a less-instructed past, and are seldom resorted to in these days.

Mares are usually served either on two following days, or on alternate days. Both plans have their advocates, but, for my part, I think the latter is by far the best. One reason for my preference lies in the fact that a mare will often be stinted on the first occasion of her being covered. When this is so, she would refuse the stallion on the third day; but, if put to him on the second, she probably would admit him, and that might tend to nullify the result already obtained. Then, again, if the first service was abortive, a renewed effort would be far more likely to prove successful on the third day, when the previous excitement had subsided, than on the second, before it had done so.

It is well known that bitches are much more likely to conceive if they are lined when just going off the heat than they are when it is coming on—that is, as soon as they will admit the dog. There are various opinions on the point with regard to mares, all of them with some show of probability in their favour. One I have faith in runs as follows: 'To cover a mare two days in succession,' says my informant, 'is simply wasting the horse. Try the seventh day after foaling, and every day after that till she comes into use. If she shows signs of doing so on the morning, say, of the ninth day, then let her be tried at about 5 o'clock p.m. on the tenth day. If she is very much at use, and can scarcely be brought away from the horse, do not let her be covered then, but try her again the next morning. If she is still much at use, have her covered on the evening of the eleventh day after foaling, and no more in that pride. Of course circumstances alter cases; but, as a rule, this plan will answer, and so two leaps are economized.' This opinion corroborates my own in substance.

After service the mare and her foal should be kept by themselves in a paddock, or with other mares similarly circumstanced. She should be removed as far as possible from the stallion, lest his neighing should excite her to turn again prematurely. This, however, she will most likely incline to do on the ninth day after the first service, and then she should be covered again. The same proceeding may be repeated until she is stinted. This she may be held to be when she has refused the horse for a month; though cases do occur when a mare will turn even after that lapse of time, and when, of course, she must be served again. Still, unless there is great reason for believing that she is not stinted, the fewer times a mare is shown the horse after her first covering the better.

About the eighth or ninth day after its birth, sometimes earlier, it is not unusual for the foal to scour. It is held that this is due to an alteration in quality in the mare's milk, which indicates that she now requires the stallion. She should be taken to him accordingly on that or the next day, for if the opportunity be lost, the condition may not recur again for some time. In the case of a mare that has foaled late, to miss the happy moment would be almost to lose the season. For if a mare cannot be covered before June, I would rather let her miss the year, and would put her to the horse early the next spring, rather than that she should foal late in May. All the same, some good horses were not born until that month, of which, I believe, *Touchstone* was one. But these are exceptional cases. It stands to reason that these late-born horses, when two-year-olds and three-year-olds, must be raced with thereby diminished chances against others having one, two, three, or even four months' advantage of them. Besides, if your mare be not served until June, and prove barren, you could hardly know that for certain until a month or so at furthest from the time she would otherwise be due to foal, if then. You would, therefore, have been prevented from putting her to the horse early in the year, thinking that she was pregnant. So, whether she bore or did not, you would either have to let her miss a year, or else go on breeding foals from her at an unfavourable time. It is better to keep her from the horse the first year, and there will be least loss of time in so doing.

Many breeders are in the habit of giving a mare a year's rest after she has bred for several seasons, not putting her to the horse at all for a year, and then in the following spring. I do not agree with this system, recognising no necessity for it in general. But when a mare slips, or bears a weakly or a dead foal, then to let her miss a year is advisable, as doing so will probably tend to strengthen her reproductive powers, and so make beneficial results apparent in the offspring she may afterwards have. Of course, many arguments may be used in favour of either system, and plenty of examples could be adduced. But I think there is no need to enlarge further on the matter.

CHAPTER XVIII.

TREATMENT OF MARE, FOAL, AND YEARLING.

An interesting study—Thorough knowledge of the subject necessary—Causes rendering mares unfit for reproduction—Turning them out too suddenly a chief one—The best system to follow.

Treatment after service and during gestation; of young mares; of mares with foals—A common cause of abortion—The same in sheep—Feeding and housing in summer and winter—Handling the foal—Attention required—Benefit of moisture to the development of the feet—Excessive moisture detrimental—Good feeding essential.

Time of foaling—Previous indications—Delivery—Abnormal presentations—The after-birth; its retention a peril—The new-born foal—Immediate treatment required by it and by the mare—Food—Suppository.

The suckling foal—How to ensure its proper development—Giving it bruised corn—Dieting the mare—The foal's mid-day meal—Time for weaning—Separation of dam and foal—Variety of food—Articles to be avoided—Condiments a mistake—Physic or no physic—My opinion and Mr. Robinson's.

The yearling—Free exercise imperatively necessary—General principles of treatment—The feet, and how to manage them—Irregular growth to be checked—Paring and rasping—Proper development of the legs depends on careful attention to the growing hoof.

THE treatment of mares and their tender offspring affords a most interesting as well as useful study.

The separation of the dam and her foal, when the time arrives for the latter to be weaned, and then its management until it grows into a yearling, and so onward, are all matters with which we who breed are familiar, and take keen pleasure in watching. Yet I think it will be readily admitted that but few are thoroughly acquainted with all the requirements and necessities of the foal and young horse; and without such knowledge it may well be asked, How are we to supply remedies for preventible diseases, to check ailments, and to correct such defects as horses are prone to when kept in captivity? These evils are probably unknown to the animals when they are running at large in their native pastures, revelling in a climate altogether suited to their well-being and comfort. But, if excellence is to be reached and maintained, it is plain we must first succeed in learning our task thoroughly, and must be prepared to meet and deal with all such contingencies as are likely to arise.

The management of mares at the mating season, and during the period of gestation, is a subject requiring more than mere passing notice. So, too, the moment of parturition, when great care, ability, and attention are absolutely necessary, must also be plainly and fully considered as we proceed.

Various causes may render mares unfit for breeding, as appears in many ways. They are often subjected to a too sudden and severe change of habit when put to the stud. They are taken, perhaps, from a stable unhealthily overheated, or stripped of the warm clothing they have customarily worn, and are immediately turned naked into a paddock, with no protection from inclement weather but an open, cold, and probably damp hovel. This must be, and undoubtedly is, a fruitful cause of disease of one kind or another. And disease, so induced, is transmitted to the offspring, while the mares themselves never recover from its effects, though they may continue breeding for years. I know that hunters feel the change they undergo whilst soiling, even in fine weather, which often enervates their whole systems and induces roaring. Why, then, should the baneful effects of exposure be felt less powerfully by mares, with their more delicate and susceptible constitutions?

The treatment of young mares, when they are first put to the stud, should be very different from this. Instead of the hovel they should have a comfortable box; they should only be allowed out in fitting weather during winter; and they should be kept in a condition best calculated to ensure health, by careful attention and liberal feeding—a state which need not be described, since all breeders are well acquainted with it, and others too. After having been served in the following spring, they may then be turned out with the other mares into the paddock,

TREATMENT OF MARE, FOAL, AND YEARLING.

and left to betake themselves to the hovel or box provided for shelter. So, duly prepared and gradually accustomed to their new life, there will no longer be the risk of inclement weather affecting their constitutions prejudicially.

After the mare has been covered, she and her foal may be led out in a small paddock a few times. This will accustom the foal to the full light, and also teach him how to follow his mother and distinguish her from other mares. Otherwise, when put with the rest of the mares, he may lose his dam among them, and, galloping about in search of her, may easily injure himself by over-exertion, or get kicked by one of the others.

The treatment of mares with foals, and of those without, during the period of gestation, is so similar that we may take them together. Both are to be managed in much the same way, except that mares with foals may be allowed a liberal supply of corn, while those without, if they have plenty of grass, do not need any until the autumn. When cold days and frosty nights come on, then they should be housed like the mares with foals, and also supplied with corn on leaving or returning to the stable; and they should not be turned out until the sun has warmed the air, or, at any rate, until signs of hoar-frost have disappeared. Abortion often results from letting mares out too soon in the morning, when the weather is cold and frosty, both in the case of mares with foals and of those without; hence the advice just given should be adopted. Sheep suffer in the same way from the like cause, and are also by it affected with the red water. Who can say how many complaints may not be developed in mares by such improper exposure?

In summer, mares with foals should be housed at night, but liberated early in the morning after having had a feed of corn. The hour should not be later than five or six o'clock, for it is then cool, and the dew on the grass is beneficial to the feet of both dam and foal; and the time of taking them in should be at sunset. In very hot weather they should be taken in for the sake of shelter during the heat of the day, being let out again in the afternoon. The need, and the time for this, must depend upon the breeder's judgment; it cannot be defined by hard and fast rules. By September, and after, the mares ought not to be let out until about nine o'clock in the morning, and must be brought back to the stable at four o'clock, or thereabouts, and be carefully attended to. These measures apply equally to young mares in foal, and to those with foals at heel.

The foal should be handled from early days. It is well to fit him with a small head-collar, with a short strap or rope depending from it. By means of this he may be readily caught, led about the box, and so be gently accustomed to restraint. His feet should be occasionally looked to; they should be picked out, the rough edges rasped smooth, while, if necessary, the toes may be shortened, on both fore and hind feet, and the heels opened. By such careful attention we may ensure good sound feet to the foal. I shall have to refer to this again presently, when dealing with the management of the yearling.

When the weather is very hot and dry, it is a necessity that the foals, with their dams, should be let out into the paddocks as early as possible; this, in order that they may derive the fullest benefit from the dew, which is often heavy enough in such seasons. Treading in the dewy grass causes the foot to expand, aids its development, and stimulates the growth of the horn. In wet weather, however, or on low-lying damp land, a contrary course must be adopted. Too much moisture will cause an excessive growth of the hoof, and will make the foot too large, as well as flat and shelly, with low heels, inducing thrush perhaps. A system of management calculated to prevent this must therefore be adopted in such cases. The mares and foals must be kept in the stable in wet weather, and, even in fine, must not be turned out until the sun has drawn up the superabundant moisture from the grass.

Mares in foal, as well as those with foals at the

side of them, should always be kept well; those without foals being housed in the latter part of September and supplied with corn. It is not necessary to fatten any of them up, like oxen being prepared for a Smithfield show; neither should they be pinched and allowed to fall poor. I like to see them in fairly good condition, neither too fat nor too lean; but I would rather see them plethoric than emaciated, in which last case there is most danger in foaling. The 'golden mean' is what we must aim at, however, especially since foals, at birth, are usually in the same condition as their dams. We see just the same thing among sheep. Hundreds of them are lost every year in lambing solely on account of being in poor condition, while the more robust get through the trying ordeal with little or no loss either to parent or offspring.

When the time for a mare to foal arrives, certain indications of the impending event give the breeder forewarning. A waxy substance forms on the tip of the teat, and some exudation of milk may shortly follow. Occasionally these signs appear some days before parturition actually occurs, as I have noticed not infrequently; but generally the mare will foal very soon after. As soon as these indications are observed, the mare should be closely watched day and night, not being let out of sight except for very short intervals. As soon as the foal has been delivered, the umbilicus, or navel cord, should be

TREATMENT OF MARE, FOAL, AND YEARLING.

tied round tightly with a piece of twine at an inch or two from its insertion in the foal's body. It may then be cut through a little below the tie, which will prevent subsequent hemorrhage, and be left to drop off of itself.

Abnormal presentations are not unfrequently met with in foaling, when the hind-quarters or legs of the foal are offered first, or it may be the head is twisted back, or some other awkwardness. The presentation of the hind-quarters need not be interfered with, as there is little or no danger in delivery that way. In other cases an experienced stud-groom is often able to set the matter right by skilful manipulation; but when he cannot do so, the veterinary surgeon should be sent for without delay, as the risk of losing either the foal or the mare, or possibly both of them, is sometimes serious and imminent.

When the mare has cleansed, the placenta, or after-birth, ought to be at once removed and buried, so that other mares may not be offended by smelling it. It sometimes happens that the cleansing does not follow so soon after the delivery of the foal as it should do. When that is the case some assistance should be given. The navel-cord, which remains protruding from the mare, may be gently but firmly pulled, every care being taken not to break it. Better still is the plan of attaching to it a small weight, so as to prevent its being drawn back into the vagina. It may then be gently pulled from time to time, and in the course of a day or two the whole after-birth will come away. If this should be retracted within the uterus there will be danger to the mare, as it will then decompose, break up, and come away piecemeal in a putrid state, and consequently everything possible should be done to prevent this occurring.

After foaling, the mare will usually clean her offspring at once, and it will shortly be able to stand. If it cannot get on to its legs, however, it should be assisted and put to the teat, and after sucking may be left for an hour or two. Meanwhile the mare should have a bran-mash, with plenty of oats in it, and a little warm water or gruel.

Some foals require to be put to the teat several times before they learn how to find it themselves; but most discover it instinctively, turning to the seat of nourishment as naturally as a duckling takes to water. After the foal has evacuated itself in both ways, it will need little more than ordinary attention to keep it in good health. If costive at first, as many are, the fact will be evident within a few hours after birth, the foal being seen vainly trying to relieve itself. In such a case a suppository, in the shape of a small tallow candle, used with care, will generally prove effective the first time, and will not need repeating.

TREATMENT OF MARE, FOAL, AND YEARLING. 247

I will go on now to describe the treatment of the foal until it is weaned and leaves the dam.

Fair size and strength are requisites of all race-horses, and in order to ensure the development of these essentials, the foal must be well kept and properly attended to, or all the trouble gone to in producing it will have been thrown away. The young creature cannot be too early induced to eat crushed or bruised oats. Generally it will do so when a few weeks old, being then observed to partake of the corn given to its dam. As soon as it does this it should have a portion set aside for itself; and, to prevent the dam eating this, when she has finished her own, she should be tied up out of reach of it for at least an hour before leaving the stable for the paddock, and the same on returning.

For the mare, I consider that two moderate feeds of corn in the day will be sufficient during summer, provided she have plenty of grass by day and hay at night. The foal will require more, however, and should have a third allowance of corn at mid-day, in fact, as much as it will eat. If mare and foal are brought in during the heat of the day, the latter can have its portion then. If they are allowed to remain in the paddock, it should have it there. A manger can be erected for the purpose, surrounded by a rail high enough to admit of the foal passing underneath it, but too low to allow of the mares doing the same. This mid-day feed is of the greatest importance to the foal, and ought never to be neglected on any consideration.

The foal should be weaned when about six months old. After that age its dam's milk does it little or no good. Moreover, if she is pregnant, continued suckling will be prejudicial to her on that account. On being weaned, foals should be put into a loose-box, two or more together, as may be convenient, and so kept for some days until they have forgotten their dams. Moreover, the separated foals and mares should be kept far enough apart to prevent them from hearing one another's neigh. For, if they can neigh the one to the other, both parents and offspring will remain restless and fretful, and their health will suffer in consequence.

After a few days' separation, the foal will lose all affection for its dam and she for it. In the matter of feeding, variety is not only permissible, but to be advocated. Too many different sorts of food can hardly be tried, but of course must be given judiciously, at various times and in small quantities. A change of food is an excellent thing for foals and yearlings, particularly as regards artificial grasses. Plenty of good hay and corn should be provided, however, as a staple. When the bowels are moved too frequently, recourse must be had to a little barley, or old split beans, mingled with the corn, while the foal is at the same time deprived of all

TREATMENT OF MARE, FOAL, AND YEARLING.

natural grass or other succulent food. As soon as well, however, it may return to the regular diet. During the same period (immediately on separation from their foals) mares may be treated in a precisely similar way.

Sometimes the mare's udders swell very much, and it then becomes necessary she should be milked by hand, once a day, two or three times. If the udder gets hard, a little goose-grease may be rubbed in once or twice, until it is soft again and in its natural state. During this time the mare should only have dry food, and not too much water.

There is no doubt that many foals are given cows' milk in almost any quantity, both before weaning and after it, and, I think, without any evil results. But I would strongly discountenance the practice of giving them beans and peas, oil-cake, and numberless kinds of condiments, which is far too prevalent nowadays. I know some excellent judges, who breed for sale, do not favour such a system, following the general lines of the treatment I have indicated, and they are usually successful. On the other hand, many undoubtedly do resort to condiments, and fail. Few of the private breeders who rear horses to race themselves use much, if any, of such unnatural food as condiments, mixtures of unknown materials. Yet these persons have, without doubt, the greatest measure of success on their side.

Now about physic. Some people give it, as a matter of course, to foals soon after weaning, whether they require it or not. For my part, I would leave well alone, having recourse to medicine only when absolutely necessary. One can always fall back upon it when requisite, and there is no need to employ such means unless they are called for. Old Mr. Robinson, of High Wycombe, used to give it, I know, and he was an authority worth following. But if I *did* think physic necessary as an adjunct to the change of weaning, whether there was any apparent call for it or not, I would administer it in the spring, just before turning the yearlings out into the paddocks; and the physic should take the form of a gentle dose of aloes, combined with the usual aromatics and other well-known concomitants. On this head I need say no more, but pass on to another.

Of all the mistakes made in the management of yearlings, that of allowing them too little scope for exercise is not the least. To put them into small paddocks totally unfitted for rearing young stock in is a sad blunder. Whether it is a foal, a yearling, or an older animal, the young horse requires plenty of room for free exercise. This is necessary in order that its natural powers may be properly developed, its frame built up, and its constitution formed and strengthened. How can this be effected if the animal only gets an hour's exercise in the day, and that in a small paddock with too many others? And, worse still, how can it be effected if the young horse is never allowed out of its box or hovel, where, possibly, it is kept in close confinement until it is sold? But no one will pretend that yearlings can be properly reared in such an injudicious way. Rather, all breeders and stud-grooms know the desirability, the necessity, of plenty of open-air exercise, and are ready to expatiate on its advantages, as well as to condemn such contrary methods as I have referred to, though only a few practise what they preach.

The treatment of yearlings should be based on the same general principles as I have recommended in the case of foals. In dry weather their feet should be carefully attended to, observing the same rules as previously. The feet are apt to get contracted, and so to become diseased; or they may grow unevenly, and will then affect the development of the legs, sometimes to such an extent as to amount to actual deformity.

If one side of the hoof gets chipped off, as it not infrequently does upon hard ground, or from striking sharply against a stone, the other side will grow redundantly, and so produce an uneven surface. From this arises deformity in the shape of the legs; the toes get turned in, or turned out; the legs assume an habitually unnatural position, perhaps becoming upright in the ankle-joints, or falling back from the knees. From high heels and short toes results a forcing forward of the leg, which tends to become permanently upright. The remedy for this is, obviously, to pare down the heels, which must, of course, be gradually done, and be commenced as soon as the evil is perceived. When the mischief takes a contrary course, and the pastern becomes too oblique, the proper thing to do is to shorten the toes and encourage the growth of horn at the heel. This will generally result beneficially.

If the hoofs are kept properly rasped, or pared with a knife, so that each foot rests flat and level on the ground, little fear need be entertained that the horse will have defective feet or misshapen legs. This is a matter, as I have before mentioned, which calls for very much more careful attention than it generally receives.

CHAPTER XIX.

TREATMENT OF THE STALLION.

Customary neglect of certain points—Common mode of treating the stallion—Pampering; want of exercise; neglect of the feet—Evil effects—How the horse ought to be kept—Exercise—Comparison with brood-mares.

Disease of the feet—Examples—My experience; what I argue from it—Rates of mortality—Causes of death—Primary reasons for such effects—Negligent stud-grooms—Treatment of the country stallion compared with that of the thoroughbred—Comparison of results—Mr. Robinson's bull—His principle.

The bowels—Constipation and diarrhœa—How to meet them—Prevention better than cure—Stallions at first indifferent to the mare; examples—This fault overcome by perseverance.

There are two particulars in which the treatment of a stallion often fails; he is not allowed sufficient exercise to preserve his health, and the state of his bowels is not efficiently watched. Neglect of either or both of these points results in the ruin or death of more horses than can be ascribed to all other causes of disease put together. Let me just describe the way in which a thoroughbred stallion is too often treated.

On his removal from the racing-stable to the stud, he is placed in a capacious loose-box. Little or nothing else is thought of for six months after, but to feed him up. He is crammed with rich food, as if he were a capon or a turkey-cock being fattened for the Christmas dinner-table, until he becomes as big and fat as it is possible to make him in the time. Nor is his subsequent treatment of a very different kind. He scarcely ever leaves his box, except in fine weather, and then only for a short walk, so that he returns to the stable little benefited by such an apology for a day's exercise. In cold or windy weather, when it snows or rains, he is a close prisoner for days, it may be for weeks together. He is carefully bedded up in nice dry straw, which is only removed when he has a fresh supply, once or twice a week.

Next, and to this I would draw especial attention, he is always shod. In consequence, the dung in which he is constantly treading packs firmly into his hoofs. It is too often suffered to remain there, perhaps for a long time, and the ultimate ruin of the horse's feet is the natural result. Ill effects may not appear speedily, but they will certainly present themselves sooner or later. Lameness is noticed, and then it is discovered that chronic and irremediable inflammation of the laminæ has been set up. Resort is at length had to removing the shoes, but the disease has now gone too far for this to be of any good, and to remedy it baffles all veterinary skill.

Now, all this is not as it should be. The stallion should be allowed a proper amount of free exercise. His feet should be looked to as carefully as those of horses in training, among which the malady is seldom known. Then there would be little or no disease among stallions; certainly no more than exists among horses to which less attention is paid in other respects. The remarks of 'Stonehenge' on this topic are very much to the purpose. He says : 'They (the stallions) are over-fed and under-exercised in proportion, so that it is no wonder they become diseased, and seldom die of old age.'

Every stallion should have, at the very least, two hours' walking and trotting exercise in the course of each day, weather permitting, except one day, Sunday, when he may do without it. On returning to the stable the dirt from his feet should be removed; they should be washed clean and rubbed inside and out with a mixture of tar and grease; and, when it is necessary to do so, they may be stopped with cow-dung. I have no hesitation in saying that such a course of treatment is the best possible preventive of inflammation of the feet, a disease which ruins so many of our thoroughbred stallions.

Why is it we do not see the same thing in brood-mares to a similar extent? Evidently because they are free to exercise themselves, constantly treading damp grass, and wear no shoes to cramp their feet. I venture to say that few people ever saw a mare with inflammation of the feet. I know I never did, except once, and that was a case due to a specific cause of another kind.

But how different it is among stallions! The first I read of who was so afflicted was *Eclipse*. 'Stonehenge' says: 'He was so lame on his feet that, on being removed from Epsom to Cannon's, in Middlesex, he was obliged to be placed in a van used for the purpose.' Many other valuable horses have been rendered useless from the same cause. When I was visiting Rawcliffe some years ago, I saw that celebrated horse, *Newminster*, so afflicted. He was so bad that he could be hardly got up to serve his mares. *Venison*, when I saw him at Broughton, was no better; and neither could stand up for more than a few minutes together any part of the day. *Young Melbourne* was another case; his feet had swelled to the size of ordinary dinner-plates. *Lord Lyon, Julius, Blair Athol*, and *John Davis* were similar sufferers; and all of these eventually succumbed to the disease.

Surely there is no need to say more, in order to prove what a fallacy it is to suppose you can keep a stallion in health under such treatment as I have described. I know that many are so kept, and what the result is I have shown. To me the wonder is not that so many suffer from the disease, but that any escape it.

The rate of mortality among our stallions, and among the best of them, too, is something incredible; yet it is a fact, nevertheless. Over-feeding and want of exercise create a great liability to every kind of disease, and deprive the animal of its natural power of resisting it. Like all other animals kept under artificial conditions—and wild ones too—stallions will die, we know. But few interested persons seem to be aware that, in spite of the best food, in spite of no expense or labour being spared to meet their requirements, and keep them in good health, there is no class of horses among which the mortality is so high as it is among our thoroughbred stallions. Appalling proofs of this could be cited in abundance.

In 'The Stud-Book,' vol. xv., appears the obituary of fifty-four stallions, out of which only eighteen died a natural death. Frequent causes assigned are :—inflammation of the lungs and breast; inflammation of the bowels, stomach, or kidneys; enlargement, rupture, or other disease of the heart, causing sudden death; fits; internal abscesses; fever in the feet; distemper; internal hemorrhage; and so forth. I venture to say that such a melancholy record of the loss of so many good horses is not to be found elsewhere.

Now, I take it that these untimely deaths are

primarily and mainly attributable to the fact of the poor creatures having been too well fed, and allowed too little exercise. That it truly is so will appear from a comparison of what I have urged with the nature of the maladies set down. Distemper and fits may be quoted as exceptions, so far as the origin of them is concerned. But even in such cases fatality is to be assigned to the plethoric and unhealthy habit of body induced by the same original faults in treatment.

Do not, however, let me be thought too sweeping in my assertions, or unjust in my animadversions. The causes I have described must not be supposed the *only* ones that operate to increase mortality among our thoroughbred stallions. Others there are, undoubtedly, as among all classes of horses.

Nor should I wish to be understood as accusing *all* stud-grooms of negligence or incapacity. Many, I know, exert every faculty they possess in the endeavour to understand the requirements of the animals in their charge, and employ every means to keep them in health. But I am sure there are also many who either do not know their business, or will not attend to it; and it is by their fault that the diseases arising therefrom are so prevalent.

Again, is it to be supposed that fat, pampered, ailing stallions should get from a mare such strong and healthy foals as she would be likely to produce by a horse habituated to a more natural life?

Evidently not. I have heard it said, on good authority, that a country stallion, led by a man riding a cob from place to place, is capable of getting more foals than one kept without exercise as I have described. Yet the country stallion often goes twenty or thirty miles during the day, and, perhaps, may serve as many as half a dozen mares at one place and another in the course of his round. Such horses are frequently known to get a hundred foals in the year. On the other hand, the pampered lord of the harem seldom gets a third of that number, although he is serving two or three mares a day during twenty weeks. Why is he not as successful as the other? The treatment of both does not materially differ, except in regard to the amount of exercise they respectively have. It appears pretty clear to me that, if we will give our stallions plenty of exercise, we shall not only keep them in better health, but shall also increase their procreative powers; while they will get more mares in foal, with less cost of strength to themselves.

Old Mr. Robinson, of Carnaby, in Yorkshire, the owner of *Melbourne*, probably did more good to the breed of long-wool sheep in that part of the country than any man before him. I mention this by way of showing that he was a practical, persevering, and able man. One day while paying a visit to him, I noticed a short-horn bull in the straw-yard. Being of an inquiring mind, I asked why he was shut in there. Mr. Robinson's son Henry, who was with me, and was himself a man of experience, replied: 'Oh, he is put there, and kept mostly on straw, to make him work'—*i.e.*, serve cows. 'If he has anything better,' added Mr. Robinson, 'he gets too fat, and will scarcely serve one.' May we not apply the same principle to horses?

Now, a word on that other point I mentioned, the consequences that result from neglecting to see that the bowels are kept in proper order. From this cause, also, many stallions are lost; and it is scarcely surprising that it should be so, when one considers how they are stuffed with dry corn and beans. During winter they have no green food; oats, hay, and beans, with an occasional linseed or bran mash, form their diet day after day and week after week, and, as I have said, in excessive quantities. Constipation is a frequent, if not always the necessary, result of this system of feeding. Then, if unnoticed or neglected, inflammation of the bowels is sure to follow, baffling all remedies too late resorted to, and resulting probably in the loss of the horse. *Young Trumpeter* met his death in this way, and so did many others.

To keep horses in good health, the state of the bowels should be watched and regulated. They should neither be constipated nor relaxed, and if they become either one or the other, remedies should be at once applied to stop the evil at its commencement. If the dung has the character of a donkey's, the horse should have bran mashes, boiled linseed, and a few crushed oats, with but little or no hay, until the hardness disappears. If it is in summer, he may have also as much green food as seems necessary to the same end. When, on the contrary, the horse is purged, he should be kept without green food, and be put on a dry diet, old oats, split beans, and hay. If this does not bring him right, some starch may be given him. But diarrhœa, when it once sets in, often assumes an obstinate character after a few days. Then the 'vet.' should be called in without delay. These, like all other ailments, are much more readily checked if taken at their inception; and, indeed, by proper attention, may be often avoided. 'Prevention is better than cure.' I think that most stallions will be benefited by having a dose of physic a week or two before the season commences.

I will now pass on to another part of the subject. Some horses, on being shown the mare for the first time, will take no notice of her. Numerous instances are on record of prolonged refusal on the part of the stallion to do his duty. *Hobgoblin* was an early case in point. He refused to serve *Roxana*, and, in consequence, she was put to the *Godolphin Arabian*. Out of my own experience I may name *Wintonian*. It was weeks before he could be got to leap a mare, and not for long after he would do that would he serve her. Several old and quiet mares were brought to him, but to no purpose. After many trials, however, he at last served a mare, and no trouble was experienced with him thenceforth. I mention this to show that a horse should not be too hastily given up as incompetent, even after so much failure as was seen in *Wintonian*.

This concludes, I think, all that it is necessary for me to say concerning the treatment of thoroughbred stallions; that is, all except what is of a special nature, and is elsewhere touched upon in the pages of this book.

CHAPTER XX.

THE HUNTER AND TROOP-HORSE.

Common origin of all kinds of horses—Hunters, hounds, and foxes of the past—Superiority of English troop-horses—Abundance of hunting sires—Thoroughbreds in North Devon; an article in *The Field*.

Employment of thoroughbred stallions for getting hunters—Modern hunters of good blood—Common-bred hunters too slow—Hounds faster than any horse.

Selection of mare and stallion—The thoroughbred cross—Shape and size—Summary of the points to be studied—Mares hunted too long; a more judicious plan—Early breeding—Early working—How the foals should be treated.

Weight-carriers—Special selection of sire and dam requisite—The Grand National Steeplechase, past and present—Nearly thoroughbred hunters the best—Anecdote of Mr. T. Ashton Smith—Inexpediency of putting hunters in harness.

UNTIL about the sixteenth century, there is no doubt that all horses were very much alike in external conformation and in physical powers. The old-fashioned charger, or 'war-horse,' as history calls him, must have been of great strength. Ancient pictures represent him moving at a gallop, though coated in heavy mail, and carrying a stalwart warrior, likewise armed *cap-à-pie*. Hunters appear to have been first used as such in the time of the first Persian Dynasty; and, probably, little modification of breed was introduced for many centuries after. At length, whether by accident or design, there sprung up a race of lighter horses, on the one hand, and of still heavier ones on the other. So, by degrees, have come all our present stock, from the huge and powerful dray-horse down to the diminutive pony.

Heavy hunters, of no breeding, and slow, were well enough in the days when hounds were no better, when you might see an old 'southern hound' motionless, whilst he gave tongue before taking up the chase. Such horses and hounds might have satisfied our forefathers, but would be despised by the hard riders of the present day. That there has been vast improvement in the speed of hounds is apparent; and this has, naturally, led to improvement in the breed of horses ridden in the hunting-field. For, if mounted on one of those hunters of the past, you would never see the hounds or view the fox after breaking cover. The fox, too, must be a speedier animal than his ancestors. This, I take it, has been brought about by the practice of introducing into one district and another foxes which were bred in distant parts. Strains have thus been crossed, and a process of natural selection, or survival of the fittest, has been going on. Had it been otherwise, we should not have had a fox left in England by this time.

I think it will be readily admitted by all conversant with the subject, that our home-bred cavalry horses, like our thoroughbreds, and, indeed, our horses of all classes, are superior to those imported. The great and gallant cavalry charge at Balaklava affords a well-remembered illustration of the speed and endurance of our troop-horses. Many other instances could be cited, but enough has been elsewhere advanced to prove the incontestable truth of this dictum.

Though many of our stallions are really not fit for the purpose their owners put them to, yet we have plenty that are. There is abundance of choice, too, to judge from advertisements, and that at a low figure. I see some in the *Racing Calendar* which will serve at two guineas a mare, and surely that is not dear. Provided the horse be suitable as a hunting sire, and be put to mares of substance and thoroughly sound, we may expect foals, in due course, that will grow into money. My previous chapters on stallions chiefly relate to race-horses, but what I have there said is not without importance in regard to the present discussion; for there is now a considerable practice of breeding from thoroughbred sires.

In an excellent article in the *Field* of November 16th, 1886, headed 'Hunting Sires,' appeared some very just and valuable remarks, which I shall now refer to. It would seem, from this paper, that there is no scarcity in North Devon of thoroughbred horses, which are at the disposal of breeders of hunting stock. Nor do I think there is actual dearth of the same in any of the shires where breeding of hunters, cavalry-horses, and other useful classes may be profitably carried on.

Lord Portsmouth, a capital sportsman, and an excellent judge of race-horses, hunters, and hounds, breeds from thoroughbred stallions, it appears, and I have no doubt has been doing so for many years past. If he has, then his hunters and carriage-horses are probably nearly full thoroughbred in blood. The late Mr. Newton-Fellowes, brother of the former earl—who once facetiously remarked to my uncle James that he wished he had been keeping sheep upon Halden Moor when his brother (the earl) was born—was fond of coaching, and drove better cattle than almost anyone else in his day. He used to say that his team was thoroughbred, and that he would not take £2,000 for it. His noble relative seems to be as fond as he was of well-bred horses.

'North Devon has been well supplied with thoroughbred stock for these last twenty years,' says the *Field's* contributor, and, from the names of thirty horses mentioned, I quite agree with him. 'Only one of these, *Gamekeeper*, had a stain in his pedigree. The farmers would not breed from him owing to the fact,' which shows how highly they appreciated thoroughbred sires. '*Comus*, a thoroughbred horse, had a hundred and forty mares put to him this season (1886), and *Omega* served a hundred and sixty.' The fact of so many entire horses being in that part of the country, and being so extensively patronized, seems to show that there can be no lack of mares there, as well as abundant demand for their produce. In fact, the writer goes on to say : ' I am told that, within these last two years, sixty young horses by *Siderolite* have been purchased for London stables, and that three big figures have been recently given for a four-year-old by *Half-and-Half.*'

All this speaks well for the breeding of hunters from thoroughbred stallions in the North of Devonshire ; and I have no doubt that if we had accounts equally reliable from other counties, we should find there is no lack of stallions and mares, kept exclusively for breeding purposes, throughout the country; not to speak of mares used also for work.

There is no doubt that, in some parts at least, crossing of mares with thoroughbred or half-bred stallions has resulted in the production of a strain very nearly full thoroughbred in blood. Some gentlemen there are who will ride nothing but animals of such a description, finding that less well-bred horses are not sufficiently fast.

In the vale, where fences and other obstructions are continually checking the hounds, one of these horses, or even a slower one, would be sufficiently speedy to keep at the tail of the hounds, if the rider had the requisite courage to ride 'straight.' Over a light and hilly country, however, no kind of horse has yet been discovered that can keep in sight of a first-rate pack on a good scenting day. Of this I am certain, for I have often tested the matter, riding first-class thoroughbred horses, the winners of sundry races. I found that, when it came to racing up and down hill, no horse living had the slightest chance with a man on his back; nor do I for a moment think he could hope to keep up with his canine competitors, even if freed from rider, saddle, and bridle. It is the knowledge of this amongst hunting men which has made them anxious to have better horses, better even than such as would have been deemed satisfactory but very few years ago.

In selecting both the mare and the stallion to breed a hunter from, it is necessary they should be well shaped, in order to produce desirable offspring. Good shape and make are as much required to give the hunter speed, as strength is needed for a cart-horse. Who, I may ask, would care to ride a weak, narrow-chested, bad-shouldered horse, whose twisted feet and legs were constantly coming into contact with each other, and threatening to bring him down? No one, I am very sure.

Such defects should be avoided by your taking care to breed only from such stallions and mares as stand well on their legs, and are of thoroughly good proportions in all other respects.

It would be useless to deny that hunters, like race-horses, have undoubtedly some admixture of foreign blood in them. On the other hand, all good hunters are now, more or less, derived from a thoroughbred cross. Generally the dam is either got by a thoroughbred horse out of a shapely half-bred mare, or, as is more frequently the case, she is the produce of a thoroughbred mare by a half-bred stallion. Many such stallions now travel the country, and serve at a low fee. Some of them are by good trotting horses. I think it is of small moment how the sire of hunters has been bred. It is more necessary to attend to his shape and size ; for all thoroughbreds are well-bred enough for the purpose of getting hunters, if they have only good and proper mares put to them.

When mating with the intent to get hunters, the following are the chief points to study :—Both the horse and mare should have good action, and should be temperate and well-mannered. They should have good heads and necks, with fine long slanting shoulders, rising at the withers. They should each have a strong broad back and ragged hips, with good propelling power to assist them over their fences. They should have strong bone, and should stand well on their legs, having flat hocks well bent under them, free from curb or any defect. They must also be sound, and have no taint of disease, hereditary or otherwise. All these points, too, ought to characterize, with equal fulness, the relatives of both the horse and the mare we select. As to size, I should prefer a mare about 15 hands and one or two inches, not more; and a stallion of about $15\frac{1}{2}$ hands high.

Generally speaking, hunting mares are not put to the stud until they have seen the best of their days. If they are fast, careful, and good jumpers, they are usually hunted as long as they can keep to the tail of the hounds through a fair run. Thus it happens that they are from fifteen to twenty years old before they are put to the stud, so that, after two or three years, they are completely worn out, and have to make room for others.

In order to breed a succession of good hunters, I would rather recommend some such plan as this: After hunting a mare a few seasons, and finding she was good, I should have her put to the horse. Say she was then nine or ten years of age; she might be expected to breed eight or ten foals as good or better than herself, before her vigour was exhausted, instead of the one or two she might have had in the former case.

If mares be put to the horse early as two-year-olds, in the months of April, May, or June, they might then be ridden quietly, and a week or two after foaling in the following year, be put to slow work on the farm, thus being made to earn their keep whilst their foals were growing into valuable animals. I know that some people think a two-year-old too young to ride. Now, thoroughbreds begin to work even a year younger, and many of them race severely also, when two-year-olds; yet we do not find they are any the worse for it in after-life, nor at all inferior to animals which have not begun so early. It is a fact I have demonstrated over and over again. I say, therefore, that two-year-old hunters and hacks may be safely ridden gently, at the back-end of the year; and that mares may be put to the horse at the same age, without fear of any injury to their constitutions, or those of their progeny, being likely to arise therefrom.

Foals of these classes of horses seldom have allowed them anything besides grass, while with their dams. Nor do they need other food, unless from ill-health, or if the mare's milk is poor. In such cases they should have a couple of feeds of corn a day, until they are weaned; afterwards, during the winter, the same, with as much good hay as they will eat, or artificial grasses when procurable. But during summer it will suffice to keep them on grass, where the land is good. This treatment should go on for the next two or three years, except that, if taken up for gentle work, they will require corn to the same extent as other working horses.

But we must also have animals available to carry heavy men in the hunting-field. To produce such it is needful to select sires and dams of more substance than those just described. For it is better to have a horse that can carry his master's weight easily, than one not up to it, which would be over-burdened and distressed before the end of the day. This would certainly be the case with a light-boned, undersized animal, if called upon to carry fifteen or sixteen stone over a heavy country. There is, of course, no more difficulty in breeding animals able to carry such a weight than there is in breeding any other. We might select a strong, well-bred, good-stepping carriage mare, about 15 hands 3 inches in height, and, by putting her to a strong, thoroughbred horse, with plenty of bone, we should probably get the animal required.

Formerly, as I have said, hunters were heavier and clumsier than they are now. The winner of the Grand National Steeplechase, in bygone days, was seldom less than 16 hands, or over that in height, and was strong in proportion. Now we see little horses, such as *The Lamb, Emblem,* and *Emblematical,* able to beat all the big ones over a stiff country; the fact being due to their better breeding, which gives them more speed and endurance. What chance would a half-bred horse have, in these days, of winning a hurdle-race against a field of thoroughbreds? None at all, I should say.

For light-weight riders you may breed hunters nearly thoroughbred. Many first-rate hunting men do so, and are carried well. I have ridden thoroughbred horses to hounds, as I have already mentioned, and, when they are steady, I like them better than others. They are quick and active at their fences, and are generally good timber-jumpers.

Lord Portsmouth, I am told, likes 'a bit of blood' as well as other hard riders in the West, and breeds mares by a thoroughbred stallion of his own choice, as I have elsewhere remarked. An anecdote is recorded of the late Mr. T. Ashton Smith, of Tedworth, Wilts. When nearly eighty years of age he said that, as he was unable to purchase hunters to suit him, he would breed his own. He commenced buying mares accordingly, and selected stallions to put them to; but whether he lived to ride any of their produce I have been unable to ascertain.

A good hunter of the present day may be made serviceable as a carriage or draught horse, but he will never afterwards be so good a riding horse as he was before. The mere fact of his having grown accustomed to throw his weight forward into the collar, when drawing a burden with outstretched neck, is enough to suggest the possibility of his hanging heavily on his rider's hands, and acquiring a tendency to stumble. Hunters and riding-horses of all kinds, therefore, should never be put into harness by those who can afford to keep them for the one purpose to which they are best adapted.

CHAPTER XXI.

THE HALF-BRED ON THE FARM.

Breeding half-bred horses remunerative for farmers—Mares can be worked as well as bred from—The preferable sort—Prices and profits—Depreciation in values of cattle, etc.—Breeding good horses a boon to farmers.

A calculation—Twenty-four mares; their produce and profit—Working and breeding—The system tried and proved successful—Additional profits.

Comparison of the ordinary shire-horse with the thoroughbred and half-bred—Respective pace—Time of ploughing an acre—Distance travelled—Gain in time, and in amount of work accomplished—Calculation of the saving and profit on a large farm.

The two sources of profit; total—Amazing result—Objections answered—Lord Lonsdale's experience; Mr. Robinson's; my own—Employment of a stable-keeper; why—Saving thereby—Have shire-horses deteriorated?—Comparison with coach-horses—Arguments—The answer.

I BELIEVE that farmers would find it remunerative to employ, for all kinds of work, strong, useful mares, half or three parts bred, putting them to thoroughbred horses; and I think this would pay better than breeding cart-horses. Some might think it necessary to keep also a few staunch cart-horses for heavy work. My own opinion is different. I think half-bred or three-quarters-bred horses would prove to have as much strength, and more pluck and stamina, than any cart-horse, and would be equal to any sort of work upon the farm. The subject is evidently one of importance to farmers, and I therefore intend to treat it with the amplitude it deserves.

Mares can be used at farm-work all the year round whilst breeding, except just a few days before and after foaling. What is more, by this practice they are kept in better health, and there is less risk of losing either them or their offspring, than when they are pampered and kept insufficiently exercised during pregnancy. Why it is so may be a mystery, but the fact is certain, nevertheless.

Scarcely any farmer or occupier of land but has some portion of it laid down in permanent pasture, enough to keep a brood-mare or two. Many holders of large farms have enough grass-land to afford provision for twenty or thirty mares, and do so keep a number of animals for working purposes. Yet few use their working mares for breeding likewise, nor, if they breed, do they do so from the right sort of horses. As I have said, it is practicable to keep mares for work, and breed from them at the same time. The sort of mare I would advise every farmer to keep would be one that is half-bred or three-quarters-bred, well shaped, and moderately large, with good shoulders. I should prefer to put her to a thoroughbred stallion, or, in default of that, to one at least as well bred as herself.

An eminent authority I have consulted on the subject says : 'If you want them (half-bred horses) to sell, when five or six years old, they must be good shire-bred ones. A gentleman I know, near Ely, sold six last month for £85 each; they were making money. But if you want horses for agricultural purposes, try a cross with a roadster or strong thoroughbred, and you will have a valuable working horse for the farm. Now and then you would get a young horse from a mare with good shoulders, a valuable animal, and worth more than any cart-horse, be he ever so good.' Good-looking produce of such animals as I have described would assuredly, when four years old, be worth in the market from £70 to £100 each, and some even more, as hunters or carriage-horses.

This view is corroborated by the remarks I quoted from the *Field*, in the last chapter, as to the produce of *Siderolite* and *Half-and-Half*. I may add that horses not so perfect in shape, though equally strong, would come in for cavalry-horses and for other purposes, realizing £40 or £50 each. Altogether the average of the prices obtained might be set down as from £70 to £80 per horse, which would have cost from £30 to £35 to raise. I think that shows a good margin of profit for the breeder.

Mr. Martin and various other eminent writers have shown how this has been done, and how it may be done again. From different starting-points they all arrive at the same conclusion, showing the accuracy of their estimates thereby. To say more would be waste of time. I am convinced the scheme is feasible, and farmers themselves, being practical men, can hardly fail to understand the bearing of these figures.

At the present time three-year-old heifers, in calf, may be purchased for £5 or £6 each, according to size and quality. Other descriptions of cattle are equally depreciated in value, often to the extent of 50 per cent. or more. It is apparent, then, that neither bullocks, sheep, nor swine can pay so well to raise as horses; and yet to the breeding of no other animal has so little attention been paid as to that of horses, except, of course, thoroughbreds for racing. As an additional inducement to farmers to engage in horse-breeding, it may be mentioned that, for farm-work, no other horses will equal those bred in the county. They are more tractable, better mannered, and of stronger constitution than any brought from a distance to the farm they are employed upon. The same truth applies, in a measure, to hunters and other sorts.

Breeding, then, from good half-bred stock should prove, I think, a boon to agriculturists generally, if to no other portion of the community. Those who sold and bought horses on a more extended scale would doubtless also be benefited; and, certainly, so would the nation at large. To illustrate my contention further, I subjoin some estimates of profit and loss.

I will take twenty-four mares as a convenient number to start with. Many large farmers keep about as many working horses; and if they keep more, or fewer, the proportions can easily be reckoned accordingly. These twenty-four mares should produce sixteen foals in the year; a two-thirds average being, as I judge from my own experience as a breeder of thoroughbreds, a fair one. Moreover, it is the average given in a statistical account in Messrs. Weatherby's 'Stud-Book,' and I see no reason why it should not apply to half-breds the same as to thoroughbreds.

Well, taking the selling price and cost of raising at the figures previously given, these sixteen foals should bring a profit of £40 apiece, as four-year-olds; that is, a total of £640. But we must allow for the loss of a mare, value £40, in two years, or £20 a year; for the service of the horse, at £2 a mare, £48; and for the loss of a foal or yearling, value £20. These deductions bring our figures of profit to £552 annually.

Of course, during the *first* three years there would be no returns. On the other hand, there would be the mares' produce available for the succeeding years. And, after the mares ceased to breed, were sold or otherwise disposed of, you would still be realizing from them for three years to come; so that the end of the undertaking would provide the just equivalent for the commencement. Of course, the same principle exactly applies to such of the offspring of the original mares as might be retained in stock.

Again, before your annual produce in foals was sold it would be realizing something, the value of which I calculate as follows: The sexes being usually pretty equally distributed, we will say you have eight colts and eight fillies. From the latter you would breed at least one year before sale. The average result would give within a fraction of five foals; but say you get four, profit on them £40 each, as before. This gives an addition of £138 to the sum obtained from the original twenty-four mares, making the total profit £690; and that is my estimate of the annual profit from breeding from twenty-four mares.

Some people may object to breed from young mares before they have been worked or sold; but I know of no reason why it should not be done. I hear, on good authority, that it is a plan adopted at Berkeley Castle by that excellent and genial sportsman Lord Fitz-Hardinge, who is acknowledged to be an admirable judge of breeding horses as well as hounds. The system has been also adopted by others; and had it not been found a good one, must certainly have been discontinued.

I might mention means by which the horse-breeding farmer could add still further to his profits. Such of his produce as were best fitted for it could be taught to jump, and might be sold as hunters. Others could be broken into harness for carriage-horses. To effect this, it would only be necessary to have on the farm a man who had lived with dealers and understood horses, to assist the farmer or his other men. A light spring van, such as jobmasters in London use for the conveyance of fodder, might be employed for breaking in the young horses to harness, single and double. By taking light loads into the neighbouring town, the horse could thus be made useful at the same time that he was being trained.

By-and-by, when the young horses had been sufficiently broken in, either as hunters or for the carriage, as the case might be, they could be shown to gentlemen in want of either sort. These would doubtless sooner give £150 or £180 to the breeder for a horse they liked, than purchase an inferior, foreign-bred animal from a dealer for £250 or £300, as they often do. There would never be any difficulty in disposing of good well-bred horses at a remunerative price; and once a reputation for breeding high-class animals had been acquired, great profits must result from the yearly sales of those raised. But it will be seen that I have not taken into consideration those larger possible profits in my previous estimates. Nor have I reckoned the value of the manure that would be formed, an item of consequence to farmers, and which might be set down against any incidental expenses I have not thought it worth while to enumerate.

Most agricultural labour in which horses are concerned consists of ploughing, harrowing, rolling, drilling, carting, and working machines. But whatever the nature of the work may be, the following remarks will equally apply to it.

Cart-horses walk at the rate of about one to one and a half miles an hour, and that pace they seldom exceed, I think, on the road, whether with a load or without it. Wherefore, if we allow that they do that on the farm all round, we shall be putting them at their utmost average of speed.

Thoroughbred horses will walk five miles an hour on the road with ease. Three-quarters or half-bred horses can do at least four miles an hour. Suppose, for the sake of argument, we put down their pace at two and a half to three miles an hour at farm-work. Such a horse would then be doing his task *within* the compass of his ability, and, moreover, would be doing it with greater ease to himself than if he were only walking at the rate of one or one and a half miles an hour. For we know that the greater the velocity the less is the resistance of opposing forces; that is, the resistance would be overcome with less difficulty by a fast than by a sluggish power. I believe that is an undisputed axiom in mechanics.

The time a pair of horses will take to plough an acre of ground must, of course, vary somewhat according to the character of the soil, according, also, to the width and depth of the furrows. But it will suffice for comparative calculation if we take, for example, the ordinary nine-by-three-inch furrow upon the light soil to which I am accustomed, as the same *ratio* must necessarily apply in all other cases.

With such a furrow upon such a soil, then, the distance travelled in ploughing an acre may be set down as twelve miles. Therefore ordinary cart-horses, walking at their highest rate of speed— one and a half miles an hour—would take eight hours to plough an acre, and that is about half an hour longer than the time usually allowed for their day's work. It is, therefore, certain that an acre is never ploughed in one day, for nothing would induce an ordinary ploughman to stay one instant longer in the field than his appointed time, as, if he did, he would be working for nothing.

Now, horses or mares that were half-bred, or more nearly thoroughbred, would walk two and a half miles an hour. They would, therefore, over the same ground, plough an acre and a quarter in the day, or half an acre more than the common cart-horses would cover. In a pecuniary sense the same difference would be found in every kind of work they might be set to on the farm. Thus, the farmer's gains must be enhanced in the same ratio, as by such horses he would gain three acres ploughed in the week. Putting it another way, the well-bred horses would plough in the year just one hundred and fifty-six acres more than the others. Reckoning the earnings of a pair at 10s. an acre, the total annual gain of the superior horses would amount to £78.

If we take a large farm, where twenty-four horses are kept, and suppose that these horses are all of the partly thoroughbred kind substituted for the ordinary lower-class animals, the gross gain to the farmer thereby, reckoned by the above standard, would amount to £936 in the year. From this, however, we must deduct extra wages paid to the men. Let us call this 2s. 6d. an acre, which would make 18s. 9d. per man per week. Calculating the present standard wage as 12s. a week, this would mean an addition of 6s. 9d. a week, for ploughing three acres more, paid to each man, leaving the farmer's profit on those three acres at 23s. 3d. for every pair of horses; that is to say, each pair of horses would earn £60 9s. in the year more than the present kind of horses earn. The whole twenty-four, therefore, would earn an increased profit of £725 8s. per annum, over and above what twenty-four horses of the common sort may be earning as things are.

Let me just explain here that I have no desire to enter upon the debated subject of wages. Whether this wage is too little, or that wage too great, is no concern of mine. I leave the question to others more interested to decide. All I want to do is to prove to farmers, by figures and calculations they can easily verify, how greatly to their advantage it would be to employ well-bred horses instead of the lower-grade animals they mostly use at present.

My former calculation of the net profit to be derived from the yearly produce of twenty-four well-bred mares was £690. I have now shown that the extra profit to be derived from the work of these mares would be £725 8s. But let us deduct from that a horsekeeper's wages, say £67 12s.; we have then left a total profit of £1,347 16s. per annum. This certainly seems an immense sum for any farmer to add to his revenue by adopting the system I advocate, yet I shall show that the estimate is still capable of augmentation.

Be it not forgotten that the small farmer must benefit, in due proportion, equally as the large one, from the same facts. I have simply selected a convenient standard on which to base my arguments. But objections will, naturally, be raised to what I have advanced, and I will now go on to deal with some delusive ideas which might be brought up against me.

First, many will say that half or three-quarters bred horses would not be heavy enough for farm-work. I am convinced that this is a fallacy. The late Lord Lonsdale used to plough with nothing else but thoroughbred brood-mares at The Links, Newmarket; and the same were used for all purposes on the farm. The late Mr. Thomas Robinson told me he had done the same for many years, and he preferred thoroughbreds to any other sort for farm-work on account of their faster walk. I have also used them myself, and found that they could do the work as well as cart-horses, and better; while, though living on the same food as the others, their consumption of it was less. To mention one case in particular, *Claverhouse* was so employed for many years, working on heavy land, and, but that he died through an accidental kick from a mare, he would probably have continued to work on the farm for years longer. Moreover, many farmers put their hunters to the plough during summer. If they can and do perform such work, therefore, what is to prevent other three-quarters-bred horses from doing the same?

In my estimates I made allowance for the wages

of a stable-keeper. My view is that every farmer ought to employ one, instead of letting the horses be looked after by the carter. The advantages of the plan will presently appear.

First, I feel sure the land would be better ploughed, by which I mean more deeply ploughed. I am an advocate for deep ploughing, though I know many farmers are not. But for these, of course, there would be less objection to the lighter description of horses, if they were satisfied with shallow ploughing. What I would say is, however, that carters and ploughmen, being also charged with the care of their teams, are too fond of seeing them big and fat, and, consequently, will not work them so hard as they would if they had not the care of them. So that, if a stable-keeper were employed, both more and better work would be got out of the teams.

Another argument in favour of employing well-bred horses in lieu of cart-horses, with a stable-keeper to look after them, is that the work would be finished in better time. The earlier in the season that the ground can be prepared, and the seed put in, the better; at any rate, in light soils. No competent and impartial person would be likely to deny that, I think. And if this desirable effect can be accomplished by the means I have indicated, it should be a matter worthy of the farmer's attentive consideration.

An important question which may be asked by agriculturists is—Are cart-horses of the present day better or worse than those of a hundred years ago? To judge from the work they do, one would be inclined to say emphatically that they are no better, if not, indeed, worse. Certainly, nothing is seen of their merits to lead us to a different conclusion. But admirers of these animals will, perhaps, meet such a startling and unwelcome assertion by urging that we are not taking a fair and proper way of regarding the matter. They will say that a horse can do no more than his driver will let him, and the driver will do no more than his master insists on his doing; that so it is now, and so it was a century ago. Be this as it may, I think it is pretty clear that no improvement has been made in the breed, if, indeed, whether from neglect or through ignorance, it is not even worse than it used to be.

We know that coach-proprietors and owners of post-horses did improve the speed of their horses, from four or five miles up to ten or twelve miles an hour. Why, then, if it could be done in the one case, may it not be done in the other? Why has not some improvement been made in the speed, at least, of agricultural horses? That is a question we are fairly entitled to ask.

When George the Fourth was looking over his race-horses at Newmarket one day, several very fine animals were pointed out to him by his trainer, Mr. Edwards. His Majesty was not satisfied with their good looks, however; but, turning to the trainer, said : 'Show me one that can *go*; that is what I want to see.'

Just so is it with regard to our cart-horses. We do not want a lot of fine, large animals, fit only for a show or for a brewer's dray ; we want horses that can *go*. We want them to be better adapted for the purposes of their existence, able to walk well, and do a good day's work.

I shall now go on to discuss another part of the subject. For it will be seen that the plan I am advocating, that of using half-bred and three-quarters-bred horses on the farm in lieu of the present sort, has a very deep and wide bearing on the whole question of agriculture in this country, affecting the well-being of the labourer, as well as the farmer. But I must, for convenience' sake, continue my arguments in a new chapter.

CHAPTER XXII.

THE HALF-BRED V. THE SHIRE-HORSE.

Continuation of the argument—How the men would be affected—Working-hours—Employers' benefits—Piece-work instead of time-work—Mr. Radcliffe and Mr. Sabin—Opinion of the latter—Differences in labourers' day; work done always the same—Extraordinary apathy of farmers—Time taken to plough an acre; instance—The eastern counties and the southern.

Waste of time—A calculation—Present system induces men to idle—My plan will stimulate them—Return of the 'good old times' for agriculturists.

Faster speed on the road also a source of profit—Calculations and comparisons—Ignorance of farmers concerning the capabilities of their horses; this exemplified—An anecdote—American surprise at our slow cart-horses—American competition; causes of its success, and the remedy—Farmers' notions as to time and work—The distance travelled in ploughing an acre worked out—Deductions from the foregoing.

THE substitution of half-bred horses for the shire-horses, and other draught-horses commonly employed on our farms, opens up so large and important a question, that I cannot leave it without fuller discussion than that presented in the last chapter. I wish to show how it must affect all classes of agriculturists, and how its general adoption would lead to the immense gain of both farmer and labourer, and prove distinctly a national advantage.

Now, with regard to the carter or ploughman, I have been told that nothing would ever induce him to accelerate his pace in the field to the rate of two and a half miles an hour. But surely such an opinion is not one to be accepted. Carters may be ignorant men, but they are not without their fair share of practical common-sense. They would see the advantage to themselves and families arising out of the new system. They would be only too glad to fall in with an alteration which would give them the opportunity of earning eighteen or nineteen shillings a week, instead of twelve, as in the old way.

The working hours for cart-horses vary a good deal in different counties, and even in different villages, as, also, do the men's wages. But, as a general thing, we may set down the nominal carter's day as from 7 a.m. to 3 p.m., less half an hour allowed for dinner. This amounts to seven hours and a half. As matter of fact, however, the carter's hours of work are really thirteen in the day. He has to be in the stable at 5 a.m., to feed and attend to his horses before going to work. On returning from the field he will have another two hours of similar employment, attending to the horses, cleaning the harness, waggons, and so forth, before he goes home. And then, again, he must

19—2

come back to the stable about 7 or 8 p.m., to see that his horses are all right for the night, which would take up another hour pretty well. Thus the working hours aggregate nearly thirteen a day, or thereabouts; besides which there is the Sunday work of attending to the horses.

Now, supposing that in the field he walked at the rate I have assumed—two and a half miles an hour, it would take him only six hours to plough his acre and a quarter, leaving one and a half hours for going to and from the field, of course an excessive allowance. Then, too, if a stable-keeper had been provided, the ploughman's work would be over for the day when he had brought back his team to the stable at 3 p.m., whilst he would have begun it at 7 instead of 5 a.m.

Surely this is very much less arduous than the existing system, with its thirteen hours' employment. Surely no carter could be so blind to his own interest as not to readily accept it. He would be receiving six shillings and ninepence a week more pay than at present, and yet be doing five hours a day less work, and no Sunday work at all. Is it conceivable that any of these men would refuse to consent to such an arrangement?

Let us see now what would be the position of the farmer under the new system. The extra expense it would entail upon him would be no more than the payment of wages to a stable-keeper and boy. This I reckon at £67 10s. per annum, and I before deducted it from the estimate of profit, leaving the total annual gain at £1,347 16s.

Now, the plan has been tried with success at Newmarket for a number of years, and both employer and employed are well satisfied with it. The late Mr. Radcliffe, once well known as 'mine host' of the Rutland Arms, took to practical farming as a man of business. It was he, I believe, who originated in its complete form the system of doing all work on the principle of paying for it by the job, or piece. Whether he actually was the first to adopt this plan or not matters little; but he certainly carried it out. He used to pay his carters and ploughmen so much an acre, and, therefore, a lad, or any competent man, could earn as much as another, eighteen or nineteen shillings a week—perhaps double what he would have got under the old system. Of course, this applies to harrowing, dragging, and all other kinds of work, just the same as to ploughing. And while the labourer is thus making better wages in shorter hours, the farmer is also advantaged; he is paying for no work but what is actually done, not, as before, paying weekly wages for work that, very often, had *not* been done.

I wonder that so easy and simple a plan has not long ago been universally adopted. It has been tried in many other cases besides those mentioned, and with complete success. I hear it is practised very commonly in America, and in some of our colonies. Mr. Delisle Hay, in his 'Brighter Britain,' talks of it as though it were the ordinary custom among the 'pioneer-farmers' and their labourers in New Zealand, as I understand it is, to a great extent, also in Australia. Why has it not obtained in England before this?

I quote some extracts from a letter received from Mr. Thomas Sabin, of Bury St. Edmunds, a gentleman lately retired from business, but formerly one of the most practical and successful agriculturists of his time. It will be seen that he largely corroborates the opinion I have expressed. He says: 'With regard to task-work, I did it for twenty-five years, and never had occasion to alter. I used to give my men for ploughing, according to weekly wages, 1s. 9d. per acre, more or less,* feed the horses in the middle of the day, and go on all day, giving 6d. an acre for drilling, 3s. 6d. per acre for filling and spreading manure, and so on. Of

---

* I assume that the meaning of this ' 1s. 9d. per acre, according to weekly wages,' is practically this: If the men received, say, 8s. or 10s. a week, the 1s. 9d. would be added to either amount for each acre ploughed. For instance, if two men receiving those wages respectively each ploughed six acres a week, one would have 18s. 6d., the other £1 0s. 6d.; but if they only ploughed four acres each, the one would have 15s. and the other 17s. per week.

course there was a lot of day-work, such as carting straw and manure to and from Newmarket, and delivering corn to the station. They drew their weekly wages, and I settled with them four times a year, after barley-sowing, turnip-sowing, wheat-sowing, and ploughing fallows. The harrowing and rolling were generally done by boys, who had to keep up to the others. I never had the least trouble to get work done, as they knew they would be paid for it, and frequently had to take 20s. each on settling-day, in addition to their weekly wages. So far things went smoothly enough. Harvest, of course, was also put out to cut, cart, and stack, at so much per acre. I had one man to look after the horses—eighteen—who lived in a cottage close by, whose duty it was to feed them and keep the harness in order, to cut chaff and otherwise provide for the horses when they were away. but never to go away from home under any pretence.'

Could any system work more satisfactorily? Both master and men were contented, and both were enriched; the men being paid for all they really earned, and nothing more or less. 'I never had the least trouble to get work done,' remarks Mr. Sabin. The conclusion is irresistibly convincing, that there was never trouble in finding men to perform work when, as he adds, 'they knew they would be paid for it.' The gist of the whole argument lies in this—pay men according to the work they do, and in any county of these islands the result will be the same: they will work as many hours, and will walk as fast, as some have been doing for twenty-five years past for Mr. Sabin, and for Mr. Radcliffe before him.

I have previously noticed that farmers do not all pay the same rate of wages. However, in my own district, the differences are not very great on the whole, I believe. Hours of labour vary likewise. Usually they are, as I have said, from 7 a.m. till 3 p.m., with half an hour allowed for dinner. But there are some farmers who keep their horses afield until 4 p.m., and some even till 5 p.m., though these allow an hour off in the middle of the day.

Now, here is an anomaly that surely wants redressing. I have consulted a number of these farmers, and they all agree in stating that, in spite of the longer hours, the extent of land ploughed in the day remains the same. I am tempted to ask, then—What good can possibly result from keeping horses and men in the field an hour or an hour and a half longer? Clearly they either idle away part of the time, or else perform the whole day's work in a disgracefully lazy style.

According to my plan, if you had horses that could walk two and a half miles an hour, and if your ploughman could make his 18s. 9d. per week, then, with longer hours, you would get more work done; for, by doing it, the man would know that he was adding to his wages. Take it which way you will, therefore, the advantage of the plan must be apparent. As at present, if an hour or an hour and a half longer time is allowed for the work, neither men nor horses earn more than they would in the ordinary seven hours and a half.

One of the most astounding features of farming is that farmers are content to go on paying for a day's work, when only half a day's work is actually accomplished. To show that three-quarters of an acre is all that is ploughed in a day, and, moreover, that it is all a farmer expects or hopes to get done in the time, however favourable the circumstances may be, I must quote one instance out of many that have come under my observation, and for the exactness of which I can absolutely vouch.

Not far from where I lived once was a three-acre field. The soil was of the lightest kind, and in all respects easy to work. One fine summer's morning, about 7 a.m., I observed four pairs of horses, and as many ploughs and men, commence work in this field. They finished ploughing it just before 3 p.m., and then left off. The ground was afterwards well cleaned, and left like a bed of ashes. About a fortnight later it was again ploughed over, the same number of horses, men, and ploughs being employed, and the same time as before being taken in doing the work.

The gentleman who owned this field was a farmer of large experience and thorough practical ability. It is evident, then, that he, in common with others, neither expected nor required his men and teams to plough more than the regulation three-quarters of an acre each in the day, even under such favourable conditions for doing more, otherwise he would certainly not have sent four pairs of horses into the field in fine weather, and in the busiest time of the year. Positive proof, this, of what I have been maintaining!

Here were eight horses doing the work that six, at the most, should have sufficed to do; walking in the field nine miles, and half a mile to and from the stable, in all nine miles and a half, for a day's work—work that might and ought to have been done in little more than half the time. The pace these horses walked at also confirms what I have said on that score. They got over nine miles and a half in seven and a half hours—a little over one mile an hour. Let it not be thought this was any solitary or exceptional instance. I have simply referred to it as one out of hundreds of a similar kind which I have noted, and upon which I have grounded my calculations and formed my opinions. The subject is truly a national one, and intimately concerns the entire community, not merely individuals. For, as the poet says, in substance, 'Delay leads to impotent and snail-paced beggary.'

One other matter should be attended to, that of feeding horses in the middle of the day. Corn can be taken to the field in nose-bags, and given to the horses during the men's dinner-time; or, if near home, they can be taken into the stable and fed there during the same period. The practice is undoubtedly a good one, and it seems to obtain pretty generally throughout Cambridgeshire, Suffolk, and sundry other counties. Mr. Sabin, whom I quoted previously, says, 'I feed at noon, and go on all day after.'

Their methods in the eastern counties compare favourably with ours in the southern. Usually, they have their men and horses working no less than one and a half hours a day longer than we do. This makes nine hours a week, during which, if the horses walk at the rate of two and a half miles an hour, a total of twenty-two and a half miles is gained. That is equal to the ploughing of nearly two acres more in the week than is done hereabouts. Broadly estimated, then, a pair of horses there will earn £50 a year more than ours. Twelve pair, therefore, would earn a surplus of £600, from which, if £156 be deducted for extra wages paid in consequence of extra time, a net profit of £444 is left. Add this to the amounts of profit and saving already shown as derivable from twenty-four well-bred mares—namely, £1,347 16s.—and we see the increment from this source raised to £1,791 16s.

Why cannot we do here at least as much as is done in other places? Is there any reason why such a lucrative system should not be as successfully carried out in one agricultural county as in another? None whatever, I am sure, except lack of energy and enterprise. It is plain enough that we could have horses which could be made to do as much work as any others elsewhere. It is obvious, too, that our men would work the longer time, and as well as others elsewhere, if it were only made worth their while to do so; for that, of course, is the root of the whole matter. As it is, do they not in time of hay and corn harvest, both horses and men, work a great deal harder and longer than at other times, and does any harm to either arise therefrom? No, certainly not. I say that you will always get plenty of men willing and eager to work, if you will but give them a just equivalent for the work you demand of them. Increase the wages in proportion as you increase the work—which is but just—and both master and men will be satisfied, and will profit accordingly.

There is another point worth attention: the actual loss incurred through waste of time, which affects the master as things are, but would only affect the men who were in fault, under the proposed system. Here are some estimates, which will forcibly illustrate the truth of the proverb, 'Time is money.'

The teams are often, nay generally, taken out of the stable a few minutes too late, and brought back to it a few minutes too soon. Say seven or eight minutes lost in either case, or a quarter of an hour in all. Ten minutes are as commonly lost in starting work, ten more at luncheon-time, and ten minutes more on leaving off work. This makes forty-five minutes a day. Let us reckon it as only three and a half hours a week. If the work that could be done in that time be now estimated at its monetary value, it will be found that this trifling loss, calculated on twelve teams, means £162 out of the farmer's pocket annually. Here is the calculation. Three and a half hours a week are equal to one hundred and eighty-two hours in the year. Horses walking two and a half miles an hour will plough an acre in five hours—not counting fractions. Therefore, the hundred and eighty-two hours are equivalent to thirty-six acres. The ploughing of these, at 10s. per acre, gives a total of £18 lost by a team in the year, or of £216 lost by twelve teams. Reduce this by deducting the ploughman's wages, 2s. 6d. an acre. £18 less £4 10s. is equal to £13 10s., or £162 on twelve teams; and that — namely, £13 10s. per team, per annum—is the farmer's *actual* loss through the waste of those disregarded minutes referred to. Now, what is the remedy for this?

I am a practical man, and have been a large farmer myself, for many years, as well as a breeder of horses, and I am sure that any farmer will agree with me there is no remedy for this waste of time under the existing conditions of work and wages. You may remonstrate with the men for being late as much as you please; it is all no use. They are always ready with some excuse or another for each particular occasion. Their watches were wrong, a horse wanted shoeing, or something of the sort. Go into the fields a little before dinner-time, and you will find them already sitting down. They will say they knew it was before the time, but they meant to go on again equally sooner. If you happen to have come upon them a little after dinner-time should be over, you will also find them seated; and then, of course, the excuse will be that they did not knock off work until after the usual time. Just so, too, they will take as long as possible to go to and from work; for the more time they can take up in this way, the less work they will have to perform. With such motives, it is no wonder they make their horses move as slowly as possible.

'Human nature is human nature,' and you cannot expect men to work properly without due incentive. As things are, the men have every inducement to waste time, to linger, and loiter, and do only half as much as they might; and this will remain the case just as long as we continue to pay them for the *time* they occupy in doing, or pretending to do, a piece of work, rather than for the *work* they do accomplish.

But whenever the system I have indicated shall be introduced, the carter or ploughman will develop quite a new nature. Being paid by the acre or by the job, as it may be, he will quickly discover that every stroke more of work that he does is so much added to his wage; while every minute wasted lessens his income. Indeed, many men, instead of wasting time, or doing only half what they ought, will rather tend to over-do things in the opposite direction, ' wanting no spur to prick the sides of their intent.' We may be sure, too, the horses will be kept at a good rate of speed, since the man will perceive that the amount of his earnings will depend largely upon that fact. As I have shown, I believe without exaggeration or unwarranted inference the farmers' profits will be very considerable by the adoption of these two principles : first, of bettering their horses, and breeding from them ; second, of paying their men by the piece or task. If anything will ever bring back 'the good old times' to English agriculturists, I am profoundly convinced it is the system I am advocating.

It now remains for me to show how the same saving and profit accrue from horses walking faster on the road, as I have shown results from their doing so in the plough. To do this is easy; let me take an example. The distance from Coombe to Salisbury Station is three and a half miles. Practical farmers about here reckon it a good day's work for a pair of horses to do this distance twice, there and back, with a load. They start at 7 a.m. on the first journey, and arrive after the second about 7 p.m. Thus twelve hours are occupied in the two trips. Let us deduct two hours for loading and unloading, and one hour for dinner-time; in all, three. We thus find that the horses travel fourteen miles in nine hours, or an inconsiderable fraction over one and a half miles an hour. That is the same rate, or much about it, which I have shown to be their speed in the field.

But suppose you had horses that walked at the rate of two and a half miles an hour—as I have shown that good half-breds will do. The two journeys would then be accomplished in something under six hours, instead of nine, exclusive of stoppages and rests. Hence the work would be done in less than two-thirds of the time, and, as I believe, with greater ease to the horses.

I have consulted various farmers, practical men, on this point. They agree in considering fourteen to sixteen miles a day in the loaded cart sufficient work for their best horses, and the time occupied to be eleven to twelve hours; and many of them are unable to do as much, except occasionally. One friend of mine in particular, living exactly eight miles from Salisbury, tells me his horses start away at half-past six in summer, and do not return until five or six in the evening, which shows the same rate of speed. I could give further corroborative evidence in abundance, if it were necessary; but surely this is enough to prove my assertion as to the rate of speed on the road of our existing cart-horses.

I have estimated the speed of half-breds at no more than two and a half miles an hour, or one mile an hour faster than cart-horses. As matter of fact, I believe they would go three miles an hour at least, as I have elsewhere shown that van-horses actually do. However, I will follow the example of George Stephenson, who, when before a committee of the House of Commons at the commencement of the railway era, said he believed locomotives could travel safely at the rate of twenty miles an hour, well aware all the time they could do much more, but fearing he should be disbelieved if he said so, or be deemed a lunatic. I might incur similar suspicion among English agriculturists were I to assert more than I have concerning what I know to be the fact, in respect to the superiority of half-bred horses over the common shire-horses.

But this difference between the two kinds of horses in their respective rates of speed has an enormous pecuniary bearing. If I have not succeeded in showing that, I have shown nothing. It is not only farmers who may be profited thereby, but also their men, whether they are paid by the acre on the land or by the mile on the road. My firm conviction is that this is a great subject, and one worthy of the most serious attention.

Without wishing to be hypercritical, however, or to appear unjust to a class of men I greatly esteem, mere love of truth obliges me to confess that too many farmers are quite ignorant as to the amount of work their horses and men actually do, can do, or should do. What their fathers did before them, that they consider sufficient for themselves to do, content to creep on in the old slow way, grumbling at, but not moved by, the changed conditions of the times, satisfied to let things go on as they used to do, and indifferent to progress.

I remember my father speaking of his father, and recalling how dissatisfied he was with the small amount of land he could get ploughed in a day—three-quarters of an acre—though on the lightest of light soil, level ground, and but few turnings. My grandfather, like some farmers of to-day, knew well enough that at least an acre ought to be ploughed in the day. But, as I have said, it was never done then, nor is it now. Why?

Here is an amusing anecdote, the truth of which I can vouch for. Mr. Cole, a farmer who lived at

Longstock, near Stockbridge, many years ago, was one day walking about his farm with a facetious friend. They noticed a plough, with horses and man, in the middle of a field, and the friend suggested it was standing still. The farmer declared it was moving, and a dispute arose and ran high between them as to which was the case. To settle the question they hit upon the expedient of getting a fold-shore, and set it up in a line with the horses heads and some conspicuous object beyond. But the ploughman now observed them, and, suspecting what they were about, became troubled in conscience and whipped up his horses, which then quickened their pace, so that the fact that they were really moving became obvious.

This incident may appear to have been of an extravagant and a preposterous kind, but, though carried out in a spirit of waggery, there is no doubt it actually occurred just as I have related. We may see examples of the same sluggishness every day of our lives. No alteration in the pace of cart-horses has been effected since our great grandfathers' days, nor do I think any improvement will be manifested as long as a carter's wages are fixed at twelve and thirteen shillings a week, and the management of the horses in the stable also left to him.

I am told by an American friend of mine, who possesses a knowledge of farming, that nothing more astonishes his countrymen when they arrive here than to see the wretched pace at which our draught-horses usually travel—and well it may. We hear a great deal of nonsense talked about free allotments of land in the United States and Canada. Free land cannot be worked for nothing, and, from various causes, is often as expensive to work as a farm here would be. In point of fact, a made farm in the States costs as much to rent as it would do here; labour is much more costly there, and so, too, is the transport of produce. The advantages of a more productive soil are thus more than counterbalanced. Yet the Americans do undersell us, despite the distance they have to bring their corn to compete against our own in English markets. The secret of their success in this really lies in the fact that they habitually get more work out of their horses, and out of their men, than British farmers dream of doing. My friend says they get double as much work from their horses, and often more than that. Here, then, is another argument for the plan I advocate. Let us follow American example in regard to our horses and men, and our farmers will find their competition is no longer to be dreaded to anything like such a degree as it is at present.

I was once told by a farmer who cultivated seven hundred acres, a shrewd, practical man, that horses had to walk twenty miles to plough an acre—a huge over-estimate, but that was his belief. He thought fifteen miles a day was enough for them, this time underrating their capabilities. Thus he was satisfied if he got three-quarters of an acre ploughed in the day, but he thought he seldom did get as much done.

Only a few weeks ago I heard another farmer—one of the 'new school,' an educated man—assert that horses travelled fourteen miles in ploughing an acre, turning a nine-inch furrow. He added to this that he thought it really did not make much difference what sort of land was ploughed, or what distance was actually travelled, as three-quarters of an acre was seldom, if ever, exceeded in the day.

These two divergent opinions seem to me to prove most amply the correctness of my previous statement respecting the ignorance prevalent among farmers as to how much their horses and men do, can do, or ought to do in the day. Evidently, therefore, they will not know when too little has been effected; nor are they in a position to insist on more being done. Of course, I admit there are numerous skilled scientific agriculturists to whom these remarks cannot be applied. But I maintain the majority are ignorant on such points; and these two instances, without mentioning others, are quite enough to show an inexcusable amount either of laxity of purpose or want of knowledge, which it would be hard to find manifested in any other line of business to a like degree.

One need not be a very proficient mathematician

in order to calculate the matter out, and to show that it does not take the horses a journey of either fourteen or twenty miles to plough an acre. In order to plough one square yard, turning a nine-inch furrow, the horses have to walk four linear yards. That must be obvious enough to the meanest capacity. Very well, then. The number of square yards in an acre is four thousand eight hundred and forty. Multiply that by four, and you will get the number of linear yards walked by the horses in ploughing an acre of land with a nine-inch furrow. Thus:

> To plough 1 sq. yd. the horses go 4 linear yds.
> To plough 4,840 sq. yds. the horses go 19,360 linear yds.
> One mile is equal to 1,760 linear yds.
> One acre is equal to 4,840 sq. yds.
> 19,360 linear yds. are equal to 11 miles.

Hence 4,840 sq. yds. (one acre) must be traversed 19,360 linear yds. (eleven miles) in ploughing.

That seems to me plain enough. But if we reckon turnings, the distance might be a trifle over eleven miles. On the other hand, surface measurement includes hedges, ditches, water-courses, banks, roads, and waste corners; so that the nominal acre might be really somewhat less. The reckoning cannot much exceed eleven miles, then, in any case. But if we put it down at twelve miles, as I have previously done, we allow an ample margin for any excess, and might also generally include therein the distance walked between the stable and the field, both ways. Therefore, if you say that the horses should walk fifteen miles in the day, they ought to plough one and a quarter acres.

Again, if you have horses capable of walking two and a half miles an hour, as I have been arguing you could have, the acre and a quarter should be ploughed in six hours, or one and a half less than the usual time allowed. If horses cannot do as much as that, and more, then the sooner they are got rid of the better, and their useless places filled up with others that can. Nor would there be the slightest difficulty to farmers in thus improving and increasing their horse-power, if they only showed a disposition to have it so.

But I must conclude this lengthy chapter, and as I am far from having exhausted the subject of which it particularly treats, I will do so with the ending so common in periodicals—'To be continued in our next.'

CHAPTER XXIII.

COMPARISON OF DRAUGHT-HORSES.

Other draught-horses *versus* the cart-horse—The coach-horse—Rates of speed—Fast and slow—The omnibus-horse; loads and distances—The post-horse—The van-horse; work performed—Good comparison—A revelation to farmers.

Dr. Johnson's ideas of progress—My reply thereto; pertinent to the question—Improvements everywhere; our agriculturists alone unprogressive—Labour and labourers—Systems in fault—Piecework encourages industry—An example from Mexico—Faults of the British farm-labourer.

Apathy of our farmers—Advice given in vain—The secret of depression—Agricultural statistics—Amazing expenditure—Immense saving possible—The means—Utilization of waste lands—Production—Comparative table of draught-horses and their several capabilities—Sorry figure made by the shire-horse—The largest and strongest draught-horse proved to be the most inefficient.

Following up the remarks included in the preceding chapter, I shall now go on to examine the capabilities of other kinds of draught-horses, such as those employed for coaching and posting, and those used in omnibuses and vans. From a consideration of their performances as compared with those of the shire, farm, or cart-horse, we shall come to the conclusion, I think, that the latter is about the sorriest specimen of the equine race we possess; or else we must admit that he is not properly worked.

Going back to the old coaching-days, we are told that, in 1775, a coach was started between London and Hereford, doing the distance either way in four days, and at the rate of three miles an hour. Another coach, at the same period, ran between London and Dover in seventeen hours, or at the rate of a little over four miles an hour. By about 1819, however, a greater rate of speed had been attained to. The coach then running between London and Exeter achieved seven miles an hour. Soon after a rate of ten miles an hour was accomplished. This I personally verified as being performed by the *Swiftsure* and the *Tallyho*, coaches which ran between Exeter and Plymouth in opposition to each other. We are also informed there were two competing coaches running between Cheltenham and Liverpool, a distance of 134 miles, which they got over in twelve and a half hours, thus travelling at the rate of nearly eleven miles an hour.

In all these cases the rate of speed cited is *inclusive* of stoppages; if due allowance was made for these, we should find that the actual speed of the horses was somewhat faster, often as much so as a mile more per hour. The distance of the stages between the change-houses was from eight to ten miles in the case of the fast coaches, and from twelve to sixteen miles in the case of the slower ones. The latter stage was no very uncommon distance, being also the usual one for post-horses; the slow coaches travelling at seven miles an hour, and post-horses doing an average of eight.

Now, our present cart-horses are only expected to draw about a quarter to an eighth of the weight these coach-horses drew, and yet we are told they cannot walk faster than one and a half miles an hour. Being so vastly inferior, then, to other breeds, one may well ask wonderingly why on earth their kind should be perpetuated!

One remarkable fact also comes to light when studying these coach-horses, which is, that those used in the slower vehicles were not any stronger or larger than those in the fast-going coaches. From this we may evidently draw the inference that big heavy horses are by no means absolutely essential for farm-work, but rather those whose pace is good. For these coach-horses were more frequently small than otherwise; yet they performed ten miles an hour, each of them drawing over a ton weight. In fact, a pair of them drew twenty-two and a half hundredweight more than a pair of cart-horses do now on the best roads, creeping at the rate of one and a half miles an hour.

Let us pass on to the modern omnibus-horse. An ordinary omnibus weighs about one ton eight hundredweight, and when it is fully loaded, weighs four tons. Each pair of horses are allotted thirteen or fourteen miles daily, and make seven miles an hour, with this weight of two tons each to draw. Yet there are horses capable of even more than this. The omnibuses plying between West Kensington and Liverpool Street have a journey of about nine miles each way. This is accomplished by one pair of horses, who go the eighteen miles, drawing from one to two tons each, only stopping for taking up and setting down passengers. It must also be remembered that the constant stopping and re-starting of the heavy vehicle makes the work a vast deal harder for the horses than if they travelled the whole way without halting at the rate of seven miles an hour.

Post-horses, though they rarely have quite so much weight to draw, travel at a faster pace, and for much longer distances. Their stages are from ten to sixteen miles; but they have to return home before their day's work is finished, which implies that they travel twenty to thirty-two miles a day. Van-horses have to do the same thing, and often cover as much as forty miles in the day.

Perhaps, by the way, these last afford a better standard of comparison than the others I have mentioned, since their pace approximates more nearly to that of cart-horses. A furniture van weighs by itself one and a half tons, when loaded six tons, and if the tail-board and roof are packed, probably several hundredweight more. This is the weight four horses are required to draw; let us say one and a half tons each. Now, a van so loaded has been known to go from Salisbury to London, eighty-one miles, and to return empty the same distance, in four consecutive days. I am indebted to Mr. H. Harfitt, an upholsterer and furniture dealer in Salisbury, for that fact. He also informs me that, taking one day with another, he reckons his horses draw the above-described load and return with the empty van, an average distance of twenty-two or twenty-three miles a day. This they do week after week, their regular pace being three miles an hour. This estimate Mr. Harfitt considers rather under than above the average.

Now, if these van-horses are capable of walking seven or eight hours a day at the uniform rate of three miles an hour, drawing one and a half tons each, one way, what on earth is to prevent cart-horses from travelling the same distance and at the same rate; considering, too, that the latter have only to drag a plough—never equal to a fourth of the weight—or a cart, about the same, or, at the worst, a loaded waggon a ton or two lighter than the van? If they are really incapable of doing this, for what possible reason do our agriculturists keep up such a wretched and inferior breed, when better abound and could be multiplied? If they can do it, on the contrary, it is a monstrous absurdity not to get the work out of them.

After the revelations made in the foregoing pages, I should hope there is no one rash enough to attempt contradiction of statements based on plain every-day facts that any farmer can easily verify for himself. I say simply this : that a farmer may, if he chooses, have horses that will walk at the rate of two and a half miles an hour, and that will plough one acre and a quarter in the day—yes, and more too. I say that by having such horses the farmer will reap very considerable advantage, and, moreover, will confer important benefits upon his employés.

I believe it was the eminent lexicographer, Dr. Johnson, who said that little had been learnt since the days of Homer, or something to that effect. Possibly this may not have been meant in the precise sense in which I understand it—in regard to labour ; but it appears to me that the application of the arts and sciences to trade and manufacture has elevated the knowledge of mankind far beyond that of the ancients, as also beyond that of Dr. Johnson and his compeers. During this century, at least, man's knowledge of himself, and of the powers he has over the forces of Nature, has progressed with enormous strides. Let us pass over older discoveries, the telescope, the microscope, steam-power and its utilization, and come down to inventions the earliest beginnings of which are within the remembrance of generations now living. Let us speak of some of these in the plainest possible way.

There is the electric telegraph, by which a message may be conveyed to any part of the kingdom for 6d., and an answer received within an hour or two at the like expense. There is the 'Macmahon Telegraph,' worked by one man, by which marvellous instrument identical printed matter can be transmitted to three thousand different stations, hundreds of miles apart, perhaps, at one and the same instant. It may well seem that thus time and space are practically annihilated. Then there is Delany's 'multiplex system of telegraphy,' by means of which ten different messages, sent by ten different operators, can be transmitted to as many different recipients upon a single wire, and this at the same time and without confusion. I understand that the Government has wisely bought up this patent. And then there is also Bell's telephone, by means of which one person may talk to another in, out, or beyond the house; in, out, or beyond the town, to the distance of fifty miles or more.

The ideas suggested by such tremendous achievements must necessarily dwarf and obscure those possessed by the ancients, and those held in Johnson's day likewise. I have referred to them only 'to point a moral and adorn a tale,' or rather an argument. I want to reason out those suggestions of mine, which so nearly, and, I think, must so potently affect the whole agricultural interest of the United Kingdom.

Whilst progressive improvements have been taking place all around us, and are still taking place, one class of people stand still where they were. Among our agricultural labourers I suppose there has been no perceptible improvement for centuries : apathetic listlessness characterizes them to-day, as it did their forefathers. Changes have occurred in the labour-market, bringing to this class some improvement of social condition, some amelioration of hardship, some softening down of asperities that are, in one form or another, the lot of all men. Yet the agricultural labourer remains the same himself, and, with seemingly dogged stupidity, bids defiance to all who would help him to attain a higher degree of comfort and usefulness. The iniquitous truck system, which was the common mode of payment for many years, has been finally abolished, and now the labourer receives his wages in hard cash. Yet he is no less necessitous now than he used to be under that system. I believe that nothing ever will really aid him until the method of paying him by the piece or task is generally adopted.

We are told that, in hot countries, where but little labour is requisite for the production of abundance of food, the labourer's natural capacity for work has sunk, with his mental powers, to that of the lowest scale of humanity. The bodily strength is impaired by want of exercise; idleness produces weakness and languor, which often lead to a shortening of life. Industry, on the contrary, braces and invigorates the constitution, adding an increase of physical strength to the frame of both the weak and the robust alike.

Dr. Hartwig, in 'The Tropical World,' says: 'In many of the silver-mines of Mexico, where the ore is conveyed to the surface by human labour, the native Indians will climb steep ladders with 240 lb. to 380 lb., performing this hard work for six hours consecutively.' Where, may I ask, is the English farm-labourer who could or would work for six hours without intermission, carrying over seventeen stone on his back to a given point, or would walk up a steep ladder with a weight exceeding that of an average sack of wheat during the same time? Are these Mexican Indians a vastly stronger race than our men; or is it that their industry, coupled with wont and use in this particular form of labour, has developed in them stronger constitutions and more muscular frames? For they evidently do what our British labourer is unable or unwilling to accomplish, who therefore remains, 'with all advantages and means to boot,' the physical inferior of the Indian.

That these things are so there can be no doubt. All speculation on the subject only brings us back to the great radical fault in the British farm-labourer : his want of energy and perseverance, and the apathetic indifference which appears to characterize him more emphatically than, perhaps, it does any other class of the community.

I am not sure, too, that the average farmer is as energetic in his own behalf as he might be ; indeed, I am pretty well convinced to the contrary. He has not moved, nor is he moving, with the advancing intelligence of the age, if, even, he has left the groove in which his forefathers were content to tread. He seems to have been but little benefited by the inventions and vast improvements in agricultural machinery. He turns to no account the full horse-power he possesses, nor other advantages that are at his disposal. Day after day, year after year, he goes on knowingly paying his labourers for work they do not perform, never seeking for means by which this might be avoided. His preventible losses are enormous in the aggregate, and, if he would save himself from them, would constitute an immense difference in his cash balance at the end of the year.

Astute reasoners have made use of argument upon argument in favour of one plan and another for assisting agriculturists in these depressed times ; but, I think, they have never brought forward the right one.

Without entering into a controversy that is baffling to political economists, it is enough for me to say that I am fully convinced the farmer has the real remedy for his present difficulties in his own hands. I think I have conclusively shown that the secret of agricultural depression here now does not lie in foreign competition and reduced prices, but in the wilful blindness of farmers to their true interests. The useless employment of money in keeping horses fit for nothing except to look at, and of wasting money unprofitably on labour, are the real causes of the present state of things.

The Agricultural Statistics of Great Britain show that in 1886 there were 22,657,494 acres of land sown with ten different sorts of seed. Let us now calculate what it costs to work this huge area, allowing each acre to have been ploughed once, and its further preparation for the seed to have been equivalent to a second ploughing. This, though an under-estimate of the actual labour involved, will give us an approximate figure of £1 per acre as the cost of the work; or a total of £22,657,494. Add to this the cost of preparing the land for green crops, assuming the expenditure to have been the same for ploughing, dragging, rolling, drilling, hoeing, carting, and spreading manure, and for cutting, harvesting, threshing, and marketing the crops, all done by manual labour and horse-power; and we arrive at a total of

£45,314,988, which is in reality below the actual expenditure.

Now apply to this result the figures and calculations I have previously given, whereby I have shown how saving could be effected in the work to the amount of one-half the present cost of it. The national gain, then, would amount to £22,657,494, taking a minimum estimate; and the whole of this vast sum, without a penny of reduction, would go straight into the pockets of the farmers and labourers of Great Britain. Surely so immense a present loss and possible gain is worth reflection, and cannot long remain neglected.

Let me add, too, that there are millions of acres in Great Britain and Ireland at present lying idle for want of capital to work them; or that are, at any rate, producing little or nothing but food for hares and rabbits, and that might be more profitably cultivated. It is evident that the huge savings effected in the manner I have described could be applied to these wasted lands. Then, in a few years, perhaps, we might hope to raise enough corn to supply our national requirements. Such an idea may seem chimerical to many, no doubt, though I would earnestly maintain it is not so extravagant as it appears on the surface. But the present work is hardly the proper medium for its full discussion.

This seems not an inappropriate place to return to

the point from which I started—namely, the utter inadequacy of the shire-horse to the needs of agriculturists. The following table will show his inferiority to other kinds of draught-horses previously alluded to for purposes of comparison:

KIND OF HORSE.	MILES PER DIEM.	MILES PER HOUR.	WEIGHT OF LOAD.
Coach-horses, four, in a fast coach	9	11	4 tons.
Coach-horses, four, in a slow coach	14	7	4 tons.
Post-horses, a pair, going 8 miles an hour over a stage of 13 miles, and returning at 6 miles an hour	26	7	1 ton.
Omnibus-horses, a pair	18	7	4 tons.
Van-horses, four	23	3	6 tons.
Waggon-horses, four, i.e., the common shire or cart horse	14	$1\frac{1}{2}$	4 tons.
Shire-horses, a pair, in the plough, and at other farm-work	9	$1\frac{1}{4}$	$3\frac{3}{4}$ cwt.

The absurdly light weight given to the plough-horses may be disputed, so it is necessary to say that I have it on the authority of one of the most eminent firms of agricultural implement makers in the country. The following is a statement I had from them: 'The draught, with a good plough, ploughing 9 inches by 6 inches in mixed soil, would be from 30 to 40 stone, or 3 to $3\frac{3}{4}$ cwt. The draught greatly depends on the sharpness of the shares and coulters, and form of breast.' But, as land is so frequently not ploughed deeper than 3 inches, the draught will be reduced exactly one-half —that is, will become only 105 lb. to 210 lb.—or half of *that* for each horse. Is it any wonder, then, that we should so often see the poor creatures with staring coats and shivering with cold, when dawdling along against this mighty draught; or that the ploughman, wrapped up in a top-coat that might resist the rigours of a Siberian winter, creeps after them as frigid and benumbed an object as the animals themselves?

The above figures show that, although cart-horses are probably the largest, heaviest, and strongest of any of those I have named, they draw by far the least weight, and move at by far the slowest pace, doing less than half as much work as the others. And the men who drive them have the easiest occupation of any class of the wage-earning community.

CHAPTER XXIV.

THE CONSEQUENCE OF THE HORSE.

Relation of the horse to agriculture—Time an estate—The agricultural labourer; his needs, powers, and capacities—The best men leave the soil; why—How to retain them—Our peasantry are literally serfs—Reasons for their laziness—Forced to the workhouse.

Means of rescue—The agitator a shameless humbug—Pictures of the possible—The labourer under the new system—What has been done can be done again—Effect on the poor-laws—Decrease of paupers—Lessening of rates—Workhouse expenditure—This digression useful—The horse the basis of reform.

My extensive disquisition on this topic of the horse in his relation to agriculture arises out of my fear that the theory I have propounded could not be explained in all its bearings in fewer words; at least, so as to be thoroughly and entirely intelligible, and so as to attract and rivet attention to it. It is better to be prolix, if perspicuity may be gained thereby, than to be brief at the risk of not clearly establishing one's arguments. To have merely asserted my conclusions, without showing by illustration and calculation how I had arrived at them, would, perhaps, have been less likely to convince than the method I have taken to bring forward all my reasonings on the subject.

'An Italian philosopher,' says Dr. Johnson, ' expressed in his motto that time was his estate; an estate, indeed, which will produce nothing without cultivation, but will always abundantly repay the labours of industry, and satisfy the most extensive desires, if no part of it be suffered to lie waste by negligence, to be over-run with noxious plants, or laid out for show rather than use.' This motto and quotation I recommend to the earnest consideration of farmers of the present day, in support of all I have said on the subject of waste of time by horses and men.

I desire to add a few words on the wants and necessities of the agricultural labourer, his powers of work, and his willingness and capacity for greater exertions under more encouraging circumstances than those he now enjoys. I may have seemed to animadvert with warmth against the labourer and the farmer in some previous remarks. But while I have sought to be candid always, it has neither been my wish nor intention to be severe or unjust. My sympathies have been enlisted on behalf of the labourer; my admiration is even due to the farmer, if for nothing else, on account of the fortitude with which he has contended against adverse times.

The most skilled artisan will always command the best wages in his particular line everywhere, and he naturally seeks the best market for the work he is skilled at. So with agricultural labourers. Their wages being some 10s. to 12s. a week, the smartest of the young men are not slow to leave their homes and seek a better subsistence elsewhere. Some become stable-men, pad-grooms, or livery-servants; some enlist in the army, and others emigrate. But if it were possible for them to earn adequate wages on the farm, these, the best men of their class usually, would surely remain among their kindred, and pass their lives amid the fields where they were born.

It would be an advantage, I think, were this so; for if the best men could be kept upon the farm, the whole class of agricultural labourers would soon be improved, and would progress further than their predecessors. Work itself would be done better, and methods improve as well as men. But, under the present evil system, any man who is somewhat abler and more energetic than the rest is sure to go elsewhere and seek more remunerative employment. Well, then, may the poet sing:

> 'Ill fares the land, to hast'ning ills a prey,
> Where wealth accumulates and men decay;
> Princes and lords may flourish, or may fade;
> A breath can make them, as a breath has made:

> But a bold peasantry, their country's pride,
> When once destroyed, can never be supplied.'

Let us rather see how it may be possible to retain our best men upon the land, and their children after them, so that we may reap the advantages which would thereby accrue. And to such advantages as I have previously alluded to, let me add that farmers would, in yet another way, feel peculiar benefit from the fact of raising their labourers to a position of self-supporting respectability. Let agricultural labourers be placed on a par with, and on a similar independent footing as, mechanics and skilled artisans of every sort, and a wonderful change would shortly manifest itself in the status of the farmers as of their men. But let me quote once more :

> 'Yet think not thus when Freedom's ills I state,
> I mean to flatter kings, or court the great.'

No ; disguise the fact as we may, our agricultural labourers, though nominally free agents, and living in a so-called free country, are still little better off than serfs. If the epithet be accounted too strong, let us at least admit them to be among the most unfortunate and wretched of human beings. Engaged weekly, one of these unhappy toilers receives the miserable stipend of 10s. to 12s. a week, sometimes even only 9s. Remember that in this I speak advisedly, however much I may regret having to do so. What is the result of this ?

Often enough, the wretched labourer works unwillingly and grudgingly. Not that he is inherently lazy or indolent, perhaps, but because he cannot but feel that he is underpaid, and therefore oppressed. In the depths of his poor ignorant mind exists a belief that the work he is expected to perform is really worth more than he gets for it. What, then, is to be looked for but that some day he should have such a difference with his employer as to lead to a separation? Then, perhaps, he fails to get work elsewhere; winter comes, and with it lessened possibilities of work. There is then no choice. The starving man, with his wife and family, are forced to go to the union. I say *forced*, because, whatever failings the agricultural labourer may have peculiar to himself, or in common with others, it is certain that he has no love for the workhouse, no desire to enter it. On the contrary, he loathes and abhors the very name of it, and only the compulsion of hunger forces him to its doors, which he enters in the same spirit as the felon enters the cell to which Justice consigns him.

The poor man enters this dismal place, the only refuge open to him, with reluctance and a sinking heart. He is separated from wife and family, and from everything he holds dear on earth. He is ' cabined, cribbed, confined,' and obliged to work for the scanty sustenance doled out to him. Is it any wonder he should sigh and long for the day of his release, that he may regain his wonted freedom, and have, at any rate, the satisfaction of earning a maintenance for his family and himself?

None may dare to chide me, surely, if I show a means, or think I can show a means, whereby these wrongs of the poor and lowly may be mitigated, or avoided altogether. Show this man how he may earn 16s. or 18s. a week, possibly more, and do you think he will not embrace the chance with joyful alacrity? Ah! surely yes! And then, perchance, there might be an end to the plaintive wails of distressed agricultural labourers—and an end, too, to the demoralizing influence of the blatant humbugs who spout and prate to them of tyrannical masters and 'bloated aristocracies'—as an end, likewise, to the subsidies these self-seekers wring from the miserables they delude.

Let the labourer have such increase of wages as I have shown he may earn, with advantage also to his employer, and he will be enabled to keep his wife and family comfortably, as well as lay by a little store against 'a rainy day.' Thus, too, would employers be saved the expense of keeping—and afterwards burying—the 'paupers' their own system has created. The institution of piece-work will do all this and more, I firmly believe, and I have shown how its introduction may be effected without loss, but gain, to the farmer. It will raise the whole status of the labourer, making him more industrious and self-reliant, a better husband, father, and member of the community. His very nature will be, must be changed.

> 'While self-dependent powers can time defy,
> As rocks resist the billows and the sky,'

so will he learn to know the value of independence, which he could not understand unless he were able to provide, by his own exertions, for the wants of himself and his family. The tap of the village 'public' will no longer have the same attraction for a man who has, in his home and family, comfort instead of wretchedness. Robuster health and added strength must also accrue to him who is better fed and less enfeebled by drink.

All this the advance in workmen's wages has already done for task-working carpenters, bricklayers, and many others who work for wages. It is only by adopting the same system of payment for work that the agricultural labourer can ever hope to give satisfaction to his employer, while at the same time increasing his own emolument, and thereby contributing to the additional comfort of his family and the wealth of the nation at large. For that the entire community will be benefited by this simple plan admits, I think, of no doubt.

One very certain effect of the substitution of piece-work for time-work among agricultural labourers will be a very startling change in the administration of the poor-laws. We know, on the highest authority, that we must always have the poor with us. But if, in rural districts, the now crowded workhouse be nearly emptied, and relief be no more asked for or required by those who are no longer paupers, the result must be greatly lessened poor-rates. And I say that this must follow the general adoption of the scheme I advocate, or else I have been writing thus to but little purpose.

It must be evident that if paupers were very much fewer in number than is the case, those there were might be far more humanely and effectively provided for. They would be mostly treated by the relieving officer in their own homes, and be maintained at less expense there than they now are in the workhouse. Thus the poor-rates would be minimized and brought within reasonable compass. The expenses of the workhouse, too, would be tremendously cut down. These are now often enormous, involving the payment of residential officials, supernumeraries, and various other persons, besides the keep of these and of the paupers, and the costly processes of the whole cumbrous system.

This is plain and sober fact, not hyperbole, and may be amply justified. Take, for example, a meeting of the Croydon Board of Guardians the other day, at which facts were brought up which will thoroughly attest my remarks. 'I see,' said the chairman, 'at the workhouse infirmary 307 inmates and 29 officers consumed 1,347 pounds of meat, 645 eggs, and 344 gallons of milk weekly. It is quite time this extraordinary expenditure was checked.' Another member of the Board, one of its finance committee, observed that he thought their system of contracting was entirely rotten, for not one of his six brother-guardians, whom he had asked, could tell him the contract price of butter, cheese, or bacon, which he found made, compared with the corresponding prices of last year, a difference of £100 to them. This loss of £100 on three items of food, as well as the preceding remarks of the chairman, may serve to show that the ratepayers' money, from want of method or economy, is not made to go so far as it might.

But I am becoming too discursive. I had it in mind merely to show how, by affording agricultural labourers the opportunity of earning better wages, as they could and would do under the task-work system, a still further benefit would result to the tenant-farmer, the rural landlord, and all others who are concerned in the payment of poor-rates. It is indisputable that these would be lessened, since many who are now and must remain paupers would be so no longer.

The whole of this chapter, as well as parts of the three preceding it, may seem to have been a digression from my main topic. But this is not so, I think. I have the subject so much at heart that I have not cared to curtail my explanations of it. I have shown how the farmer may be benefited, how the labourer may be placed in a better position, and thus how not merely the agricultural community, but the whole nation itself, must largely profit. And the basis upon which the whole fabric of my argument rests is the improvement of the breed of horses used in agriculture. It is to the horse—to his efficiency, capabilities, and perfection—that we must look for the redemption of the agricultural labourer from his present deplorable condition, and for the extrication of the British farmer from the many enormous difficulties which now beset him.

CHAPTER XXV.

DRAUGHT-HORSES AND TROTTERS.

Farm-horses—Good thoroughbred sires for hunters—Selection—Pedigree not so essential as shape, etc.—Qualities of ancestors to be studied—Good make very necessary—Mating cart-horses.

Various draught-horses—The dray-horse—The Cleveland Bay—The Suffolk Punch—The Devonshire Pack-horse—New strains—Breeding and use of all these—Intermixture of old local breeds—Gradual improvement.

The American trotting-horse—Mr. Hiram Woodruff's book about it—*Messenger*, the parent sire of the breed—His son *Top-gallant*; an extraordinary horse—Anecdote of his speed—A race in the dark—Another 'yarn'—Extraordinary fecundity of *Messenger* and his offspring—*Flora Temple* and *Dexter*; their feats—Mode of rearing trotters—My criticism on Mr. Woodruff's views—Physic and diet.

From what has been advanced in preceding chapters, it will be seen that we are largely interested in trying to discover the best description of horses to be used in farm-work generally. It has been plainly shown that great advantage must ensue to both employer and employed, as well as to the nation at large, from getting horses for the plough and cart capable of walking three miles an hour, and doing it as easily as the present kind can do half the rate. Having found a preferable kind, it must also become matter for study to ascertain the best way of breeding such horses at the least possible expense. That done, and our grand object will have been practically attained.

In further reference, too, to the breeding of hunters, I may say that I prefer small stallions. Some such that might be named as being likely to improve the breed of hunters, if only mated with mares suitable to them in every way, are—*Lord Ronald, Master Kildare, Mr. Winkle,* and *John Davis.* Then, too, we have *Perion,* which, like *Dear Me,* is described as being 'a very little horse, but one of the most useful hunting-sires ever seen, and his daughters are said to be equally good at the stud.' These stallions are, to my thinking, superior in a very high degree to any imported horses, or indeed to any larger horses either, for the purpose of getting hunting-stock.

Now another point. I have laid great stress on the attention which should be paid to the pedigree of thoroughbred sires when selecting for the purpose of breeding thoroughbreds. But this is not so necessary when our aim is merely to get good half-bred stock. All thoroughbred horses may be considered good enough to improve inferior stock, so far as that goes. But in selecting for this purpose, we ought to study shape and make, which are of infinitely more consequence in this instance, as well as soundness, constitution, size, temper, and action. Scrupulous attention should be paid to each and all of these points, both in the stallions themselves, and in their parents on both sides. And, indeed, it would be well to choose horses whose ancestors, for several generations, were well qualified in some, at least, of these particulars. Such only can be best recommended to the breeder.

These admirable qualities, I say, should be looked for on the side of both parents, as elsewhere advised. You are more likely to get the stock you want if the desirable qualities are found on *both* sides, than if you get, say, speed on one side only, and stamina on the other only. It is advisable to go back some generations in order to avoid, where possible, a soft strain affecting your choice. For, where you find one good stayer, you will find a hundred that are more or less jady, and the produce is more likely to inherit from the soft than from the game strain, on the principle, no doubt, of its being the preponderating one. It is precisely the same with draught-horses. Weakly-natured animals are always more easily bred than such as have robust constitutions and are honest and good workers, if care be not taken to breed from parents possessing the desirable qualifications here mentioned.

When selecting for the purpose of breeding

draught-horses, too, of whatever sort, the shape and make need most careful attention. Good shoulders are essential as much as well-set legs and feet. The very best procurable should always be selected to breed from. Mr. Gilbey, speaking of the shire-horse, very properly observes: 'A compact, well-formed cart-horse will move a given weight with far greater despatch and less chance of injury to his powers than can one whose shoulders are defective, whose loins are bad, and whose legs are ill-formed.' And so it is in breeding any description of draught-horse.

I have said nothing with regard to barrenness and slipping in other kinds of mares than the thoroughbred. Nor do I think there is any need that I should, for it would be only repetition. What has been said in the chapter devoted to that subject applies equally, at least in all general details, to all brood-mares, of whatever description, quite as much as to thoroughbred mares in particular.

We may now go on to consider the sundry kinds of draught-horses, light and heavy. The largest sort is that known most commonly as the dray-horse. This huge magnificent animal stands from eighteen to even twenty hands high, and is big and strong in full proportion to its great height. Such horses are mostly used for very heavy work, notably to draw brewers' drays; as also vehicles conveying

22—2

contractors' materials, large timber, stone blocks, iron girders, and the like, as well as vans transporting heavy goods of every kind. Horses used for such purposes must have weight, and this is a great desideratum in breeding them. For such work, the larger and heavier the horse is, the greater is his usefulness, and, consequently, the greater his value becomes. Horses of this kind may be bred from Cleveland Bays, from large shire-horses, and such-like descriptions. They are to be treated and reared in much the same way as carriage-horses.

The Suffolk Punch, the Cleveland Bay, the Devonshire Pack-horse, and various other once noted breeds, have become in time so much intermixed with other classes that it is with the greatest difficulty we can find a pure-bred example of any one of these specified kinds. The largest description of shire-horse may be traced to descent from the Cleveland Bay. This, again, certainly owes its origin to the old black cart-horse of the country in by-gone times. But these are now probably lighter and less bulky than they used to be, due to admixture, perhaps, with thoroughbred blood.

The Suffolk Punch was originally obtained by putting Norman stallions to the mares then bred in the county, and hence the name which continues to attach to descendants of that cross. The breed is now not so often met with as formerly either in Suffolk or in the neighbouring counties of Norfolk or Cambridge. Even here, in many districts, it is often not to be seen at all, having been supplanted by a new breed, one consisting of chestnut horses of a similar general type to the old Suffolk Punch, but not of the pure breed or of equally excellent quality in its full development. Much the same thing obtains in regard to the old Devonshire Pack-horse. I think that breed must have been crossed out of existence altogether. It seems to be totally unknown at the present day; at any rate, I have been unable to discover any purebred specimen of it yet left.

The heavy Cleveland Bays were formerly much used for breeding hunters and carriage-horses, but they have long since had to give way to animals of a lighter sort. For a good many generations back now—some six or seven very probably—they have been crossed with thoroughbred horses, and with the bay coach-horses of the country, which last cross still gives size and bone. The pure blood of the Cleveland Bay is fast disappearing, however, if, indeed, it has not been totally lost ere this, in spite of occasional re-introductions purporting to be pure, but to be very doubtfully accepted as being really such.

When the Cleveland stallions had come to be used too long, and the resulting stock had become too large and unwieldy for any ordinary purposes, a fresh strain was introduced. They were crossed with the Chapman type, a class the value of which is well known to and recognised by breeders. This gave rise to another class, one not dissimilar to that largely bred and used in Lincolnshire some time ago, though rather heavier. Devonshire and other parts of the country used to have their respective indigenous breeds; but, from one cause and another, these are now to be met with in much less numbers than formerly. Greater and quicker facilities of communication have probably had much to do with it, bringing about interchange and intermixture to an extent that has been fatal, or nearly so, to local strains.

In earlier and more primitive times there could not have been anything like the present possibilities of choice for mating sires and dams. The war-horses of the past were probably what we should term useful, though heavy, riding horses. They were most likely got by putting the best draught-mares of the period to stallions of the same class, or of one that did not differ very greatly from it in general character and quality, such as they had in those times. Very possibly they also selected mares of other kinds that were accustomed to slow work, for strength more than speed was what they chiefly sought to obtain; and probably they did obtain it by the methods indicated for breeding their horses.

However, just as the breeds of fox and stag hounds have increased in size, and become of better speed and more endurance than those employed by our ancestors, so have horses improved. Hunters have had to be bred equal to the better kind of hounds, and so stallions and mares of a lighter description came into vogue for riding purposes, while the heavier sorts were relegated more entirely to drawing. Probably fox-hunting and stag-hunting have had more to do than anything else with the production of horses of a speedy kind; strength and size being no longer considered so desirable as swiftness and endurance. Thus came in the thoroughbred cross, and the most fashionable hunters and riding-horses became such as they now are, that is to say, horses got by thoroughbred stallions out of half-bred mares, or out of such well-bred mares as had been hunted and proved good in speed and stamina.

Now a few remarks respecting the American trotting-horse. Mr. Hiram Woodruff's work on the subject, published in 1876, is in my estimation a very excellent and able treatise, and I shall accordingly quote a few passages from it. Mr. Woodruff deals with the management of the trotting-horse from its birth onwards, and describes the arts of riding and driving with all the graphic piquancy with which Yankee idioms can invest the topic.

His observations on breeding and rearing are not very copious, yet they are often worthy of record; and in many instances his views coincide with those I have expressed in this and other works.

The present breed of horses in the United States is vaunted by its possessors as being the best in the world. As trotters, at any rate, they may be safely accorded the honour of being the fastest in that particular pace. The strain is said to have originated in stock got by an English thoroughbred horse, *Messenger*, who was imported into America in the latter part of last century. Mr. Woodruff describes the advent of this famous sire in the following words: 'When the old grey came charging down the gang-plank of the ship which brought him over, the value of not less than one hundred millions of dollars struck our own soil.'

Messenger was the sire of *Top-gallant*, which horse was not known as a trotter until he was fourteen years old, and then ran most successfully until he attained the age of twenty-four. 'In fact,' says Mr. Woodruff, 'in some respects he was the most extraordinary horse that ever came under my observation.' From what he further goes on to relate of this animal, one may conclude that he truly must have been the most extraordinary animal that ever came under the observation of anyone. Says Mr. Woodruff, '*Top-gallant* went so fast in a match that the people could recognise neither him nor me, and remained in doubt what it was that had come like a flash through the crowd and won.'

We must not say that this stirring account of one of *Top-gallant's* feats is too highly coloured, but it is certainly very astonishing. Then follows the description of a match that was run in a severe storm, and which is equally amazing. 'Neither jockey,' he says, ' could see the other nor hear the wheels nor the stride of the horses, by reason of the wind and rain. The drivers had to call to one another apparently, in order to prevent disaster. '"Look out, Hiram!" Spicer would say, "or we shall be into each other." A few strides further I would sing out, "Take care, Ginger; you must be close to me!"' The tale is altogether marvellously grand, and the incidents graphically described. I could wish to reproduce it in full, did space permit, in the rich dress of the original dialect.

However, we should indeed be wanting in courtesy and in gratitude did we attempt to deprive Brother Jonathan of the credit due to him for his ingenious inventive faculties, as well as for the inimitable manner in which he is able to describe matters for our benefit and relate tales for our amusement. Yet it is only fair to say that there are, even on this side of the 'big drink,' many things related which would strike Brother Jonathan as being quite as comically surprising as his narrations often appear to us. Here is an instance.

It is related by 'Druid,' in his 'Scott and Sebright,' that '*Attila's* trial at Newmarket at two years old was quite in the dark, and Colonel Peel's *Hardinge* and *Sir Harry* had just tried him half a mile on the Limekiln Hill, when the renowned J. B. arrived with a lantern to reconnoitre. John Scott could not see the horses, but he knew from *Attila's* peculiarly quick and delicate step that he was coming away in front, two hundred yards before they finished.'

It seems to me this 'yarn' is not far behind the Yankee's in respect of the extraordinary nature of the feat it describes. One would like to have asked how John Scott knew the exact distance given, who placed the horses or who was the judge, if the last was necessary at all on such a sombre occasion. But we are left considerably in the dark ourselves, for the horses were not all placed, nor is any account given of the way in which they finished, nor of what precautionary measures were taken by Mr. John Scott to avoid being run over during this race in the dark. No; when it comes to 'spinning yarns' and telling tales of the marvellous, there is not much to choose between the Britisher and the Yankee, after all, but both must bow before the still mightier genius of Munchausen.

To return to *Messenger*, I will confine myself to saying that he was one of the most successful horses that ever lived, according to the description given of him. He must have been the oldest, too, as he is said to have seen forty-four years. Nearly all the best trotters of that day were got by him, while his descendants for several generations, stallions and mares, were equally famous at the stud. Mr. Woodruff says: 'He covered for twenty years, and must have been the sire of a thousand horses; his sons were as long-lived and as thoroughly employed in the work of increase as himself, and his grandsons continued to possess the same qualities and peculiar gifts of their ancestors, even to colour.'

Flora Temple, a bay mare, and *Dexter*, were both descended from *Messenger*. The former trotted in harness two miles in 4 min. 50½ sec., the latter did the same distance in 4 min. 51 sec., the mare thus making just the best time. She was but little larger than a good-sized pony, and *Dexter* was not much bigger either. These feats deserve to be considered extraordinary.

Alluding to the mode of rearing the trotting-horse, Mr. Woodruff advises us 'not to overdo him when young with oats.' I should think it would be an impossibility to 'overdo him with oats,' if he only were allowed sufficient room and liberty for exercise. 'Avoid Indian corn,' he further says, 'give no physic; it is only to cure sickness, the prevention of which belongs to diet, careful attention, and judicious treatment.' In all this I have no doubt Mr. Woodruff's opinion is sound. Physic is much too often administered to horses when in health, to their detriment; and feeding receives too little attention, both causes of many disasters.

Our author goes on to say that 'early maturity means speedy decay; long trotting work makes horses slow, and too much walking exercise is bad.' Exactly what he means by this advice I am unable to say; it is altogether too indefinite and vague. For instance, does 'long trotting' imply trotting for five or six hours at a stretch, or attempting too great a distance in a short time, or continuing to trot races too long in one season, or in several seasons? The advice, it is evident, may convey totally different impressions to one person and another. So, too, the remark that 'too much walking exercise is bad' requires explanation. Why and how is it bad, and what is too much? These vaguenesses leave us in doubt, and considerably detract from the merits of a book which, as I have said, is not without value on its particular subject.

It will be noticed that I have, throughout this work, cited the opinions of numerous breeders and other authorities either in contradiction or corroboration of my own. This, I think, is necessary in a work of this kind, for it is by no means my desire to 'lay down the law,' nor pose as a self-constituted authority from whose views there is no appeal.

Mr. Woodruff, speaking of the American trotter in its early youth, tells us not to overdo the foal with oats. His views seem to receive some support from a gentleman upon whose authority I generally place great reliance. The latter tells me 'no foal should have more than one and a half feeds of oats up to the 1st of January, given at three times daily;' that is, a thoroughbred born in or after the previous January, or, in other words, any foal until it is nearly a year old. 'After that,' proceeds my informant, 'two and a half feeds at four times in the day, with daily paddock walking exercise according to the weather. . . . Whereas many have been glutted before the trainers get them, and they go all to pieces, till they won't feed at all.'

There is much truth in this statement, doubtless; but whether the number of feeds mentioned be sufficient or not entirely depends on the size and weight of each. One so-called 'feed' may consist of a quart or three pints measure, and another may mean double the quantity. Again, oats that weighed only 38 lb. per bushel would be inferior to such as weighed 44 lb. per bushel. No horse could eat the same quantity of the heavier as he would of the lighter. Yearlings are, no doubt, often too much pampered, and possibly surfeited with prepared food until they will eat nothing at all, as above remarked. But I have fully discussed the subject in a previous chapter.

CHAPTER XXVI.

THE GALLOWAY, PONY, AND ASS.

Small kinds of horses—Derivation from a common stock—The galloway—The pony—Feats of endurance; my father's boyish experience; *Sir Teddy*—Origin of small breeds—Cold climates and poor living; this questioned—Other causes may have operated—Arguments—Natural selection—Giants and dwarfs obtainable from a common source.

The ass—Various kinds—Syrian and Spanish asses—Crosses between equine and asinine species; kinship yet distinction—Swiftness of asses—The onager or kiang—Improvement of the English ass — The 'coster's moke' — Endurance — Country donkeys—Reasons for poorness of the breed—The English ass the most inferior of his kind—Selection and treatment called for—Comparisons—Great improvement possible—A suggestion to landlords and others—What has been done for horses might be done for asses.

I SHALL now have something to say concerning such classes of horses as hacks, galloways, and ponies. Some people suppose that the galloway, or, as it is sometimes called, the racing-pony, is a distinct breed. I do not know of any sufficient authority for this statement, however, nor have I found that galloways are anywhere more exclusively bred than hacks or ponies. The one name is as general in its application as either of the others. Animals of thirteen to fourteen hands in height get called hacks, galloways, cobs, pads, roadsters, and so forth, according to their build, or to the use that is made of them, the galloway being the smallest, and merging downwards into the pony, which is of less height, sometimes being very diminutive.

All these descriptions of small horses may be derived from the self-same ancestors. They are usually got, when of good class, by a pony, whose sire was a thoroughbred horse, or by a galloway, also well-bred, out of a small mare, one such as would herself be styled a galloway or pony, probably. When of good shape, and well-bred, these galloways will do an astonishing amount of work, and their powers of endurance are truly prodigious, when they are not over-weighted. They are found to be equally useful in harness as with the saddle, and have frequently been known to travel a greater number of miles in a day than horses much larger than themselves could manage to accomplish.

Less in size than the galloway, being more or less below thirteen hands high, is the pony. Perhaps no sort or kind of the equine race are capable of performing greater feats of endurance than these plucky little animals. As I have related in a former work ('The Race-horse in Training'), my father and Mr. Montgomery Dilby were once participators in a feat of pony-riding that is worthy of record. As boys, they one day started from Exeter, after the races, mounted on small ponies, and rode to Stockbridge, a distance of 107 miles. They must have left Exeter about 5 p.m., and they arrived at Stockbridge at 7 a.m. Thus the rate at which they had ridden on their gallant little steeds may be set down as having been eight miles an hour, since they had made 107 miles in some 14 hours; and this would have been inclusive of all stoppages, of course. Few horses in that day were capable of performing such a feat, and not many now would be able to accomplish it.

Another astounding feat of much the same kind has been recorded of a pony called *Sir Teddy*. Strange to say, it was performed, in part at least, over the same ground as in the case just related. *Sir Teddy* was twelve hands high, and what he did was to race the mail-coach from London to Exeter, and beat it by 59 minutes. This extraordinary little animal had no weight to carry, being led between two other horses, from stage to stage, all the way. He got over the entire distance of 172 miles in 23 hours 28 minutes, or at the rate of about $7\frac{1}{3}$ miles an hour, including stoppages. It is singular that such gameness, as well as physical power of endurance, should be so much more frequently evinced by the smallest sort of horses than, comparatively speaking, the larger. Yet that it is so is an indisputable fact.

The origin of some of the small, and even diminutive, breeds, such as the Shetland, Welsh, New Forest, and Dartmoor ponies, has often been ascribed to the poverty of their food, the poorness of the hill pastures on which they are reared, and the damp and biting cold of the climates which prevail in the districts they inhabit. No doubt these circumstances have much to do with the diminution in the size of animals. We see, for example, in Iceland that horses, sheep, and cattle are all of very small size, and there seems a general tendency among such animals to become less the further north we go, as may be seen also in the north of Scotland and of Scandinavia. Cold, poorness of food, and a scanty supply of it, are, therefore, held to induce deterioration of size.

But I am not at all satisfied that cause and effect are here rightly ascribed to each other; or, I should say, that the cause assigned has been sufficient in itself to produce the diminution of size generally supposed to be due to it. Among the wild animals of the extreme north of America and of Asia are some which show just the opposite character. There might be mentioned species of deer, bears, and the weasel tribe, inhabiting the coldest and most inhospitable regions, which are actually much larger than their nearest congeners in warmer climates. Judging from specimens in the Natural History Museum at Kensington, the musk-ox, a purely graminivorous beast, which is found far within the Arctic Circle, appears to be almost, if not quite, as bulky as the bison or buffalo fed upon the rich prairie-grass of the American plains.

There seems some ground, then, for doubting whether cold and poverty of food are, at least, the sole causes which have given rise to the small breeds of horses, sheep, and cattle found in the extreme north of Europe. And this view is strengthened when we also look at the fact that these small breeds are distinguished by a more considerable share of physical endurance and hardy qualities than the larger of their kind. Is it not more probable that these little sorts have arisen by natural selection, they alone, of all their species, possessing the qualities which have enabled them to survive in the situations in which they are found? At any rate, the question, whatever be the proper answer to it, brings us back to the fact that the smaller kinds of horse are relatively stronger in constitution, of greater physical capacity, and more enduring, than are the larger kinds.

It is certain that it only requires a given number of years to admit of the production, from the same original parents, of the largest and the smallest types of a species. I have already spoken of the gigantic Cleveland Bays, of their being put to thoroughbreds, and of a lighter cross resulting. In like manner, I have no doubt, ponies and galloways have got thoroughbred blood among them, and perhaps such fine specimens as *Sir Teddy* owe, or have owed, their courage and indomitable pluck to the fact. Most classes of horses, nowadays, are more or less frequently crossed with thoroughbreds.

Something may now be said of another member of the equine race—the ass; and he must be the last on which I can dilate. For it would evidently be useless to say anything here of the zebra and quagga, they not being in a state of domestication anywhere that I know of; and speculation on their probable merits or demerits if subjugated would be but assumption, and, therefore, of no practical service.

The domestic ass is widely spread over many parts of the habitable globe, and, to the poorer classes of society, is of even more service than the horse. In the Hebrew Scriptures we read that the ass was known to the ancients, and was used by the patriarchs. Most likely his employment by man was anterior to that of the horse. His origin may have been in Central Asia, or was, perhaps, African, for the present domestic species greatly resemble the wild ass of Abyssinia and neighbouring parts. But in the North of Persia, and in Tartary, Thibet, etc., exist numbers of Onagers (*Equus hemionus*), a species of wild ass.

To describe the colour, peculiarities, and habits of so well known and useful an animal would be altogether superfluous. Everybody knows the common donkey, and is more or less familiar with his general character. But, probably, not so many are acquainted with the fact that there exist breeds of asses as immeasurably superior to the ordinary English donkey as is the thoroughbred horse to the most inferior cart-horse. The Syrian ass, both wild and domesticated, is very much larger than any other of its species, and differs from them also in conformation and external appearance, having more the character of the horse. Very possibly this may have been the animal referred to when asses are mentioned in the Bible. The Spanish ass, too, though nearer akin to our own breed, has been bred with care and selection, so that it is often as big as our galloways. This breed was introduced into South America by the Spaniards, and is now very numerous there.

Cases of fertile union have been recorded between the horse and the zebra, the horse and the quagga, the horse and the onager, the horse and the common ass, the ass and the zebra, etc., which prove indubitably that all belong to the same generic race—the *Equidæ*. This genus is divided into two distinct types: the horse one; the ass, zebra, and quagga forming the other. Both types are distinguished from all other animals by an undivided hoof, a simple stomach, and the teats of the female being situated upon the pubes.

Despite this kinship between the horse and the ass, however, it has been conclusively established that the two belong to distinct and different families, and not that one has, in course of ages, departed from the form of the other. 'There is that inseparable line drawn, that barrier between them, which Nature provides for the perfection and preservation of her productions—their mutual offspring, the mule, being incapable of reproducing its kind.'

I find it stated in the 'Encyclopædia Britannica' that asses 'are remarkably swift, having been known to outstrip the fleetest horse in speed.' This must be, I think, an exaggeration. If it applies to any kind of ass, it must be to the onager of Central Asia, which is probably the speediest of the ass family. But the statement lacks confirmation, and, so far as our present knowledge goes, it is impossible to receive it as anything else but an extravagant assertion.

A kiang, or onager (*Equus hemionus*), was brought from Thibet some years ago by Major W. E. Hay, F.L.S., who presented it to the Zoological Society, in whose well-known gardens it may possibly still be living. This animal became very tame and docile, and exhibited all the hardihood of the common donkey. I do not know if any attempt has been made to cross it with the ass or horse, however. Major Hay describes the wild kiangs as ascending and descending rocky hills with wonderful rapidity, and without ever losing their footing.

I believe that our domestic asses might be very greatly improved, so as to become available for very much more general use than at present. There seems, indeed, no sort of reason why we should not have asses as large and fine as they are bred in Spain and South America. But, without going as far as this, it would surely benefit a large number of poor folks if the sort we have were made better.

Among the costermongers of London, who generally take a great deal of care of their useful 'mokes,' I have seen a donkey that could trot his seven or eight miles an hour with a heavy load behind him, and would continue travelling at about this pace for several consecutive hours. Surprising as such a performance may be, it is quite eclipsed by a feat I find on record. On the 30th of May, 1761, Thomas Dale rode an ass 100 miles in twenty-two hours and thirty minutes. This is at the rate of four miles and four-ninths an hour. Allowing for several stoppages which he must have made, amounting altogether to four or five hours, probably, the average speed must have really been about six miles an hour.

In the country, however, in contradistinction to the Metropolis, the poor donkeys are taken much less care of—in fact, are very often hardly cared for at all. Among them, therefore, it is a very rare circumstance to find one that can trot much, or go faster than a walk, at a pace of about a mile to a mile and a half per hour. 'Like a donkey's gallop, short and sweet,' is an old saying, and one fully borne out by these country asses; at least, by such of them as are able to gallop at all.

One circumstance militating against improvement in the breed of donkeys is that they are so few in number in comparison with our horses. The selection of parent-stock is, apparently, never attended to in the least. All sorts of asses are mated without the slightest regard to their respective qualities or imperfections, and are bred from indiscriminately. Nothing is done, therefore, to make up for those processes of natural selection which would modify the race in a wild state. Belonging, as they mostly do, to the least opulent sections of the community, neither expense nor judgment regulate the mating of asses. A she-ass is put to the jack-ass of a neighbour, the nearest there happens to be to the owner's residence, if he do not possess both sexes himself. The produce is brought up on the very worst of food; it is badly housed likewise, if, indeed, it is so favourably circumstanced as to be housed at all. Such other wants as, in common with the horse, the donkey has, are entirely and absolutely denied to it.

Existing under such unfavourable circumstances, it is little wonder that our English donkey is about the most degenerate specimen of the asinine race that the world can produce. That there has been no improvement in our breed is a positive certainty. That there has been gradual deterioration is far more likely, if it be not so certain. And here we see a most surprising fact, and one that is little to our credit as a nation. For, while we have improved our horses, until the English thoroughbred has become undoubtedly without a superior anywhere, we have, on the other hand, achieved the distinction of producing the poorest and worst donkeys known to exist.

The subject is not altogether a new one, I am aware. Here and there individuals have made some spasmodic attempts to introduce better asses into the country. But either they have not gone the right way to work, or have tired of the undertaking before any satisfactory result had been obtained. What is wanted is more sustained and general effort, and by-and-bye we shall see a noticeable improvement in the form and size of our asses.

There is no doubt that by proper and adequate treatment, and by careful selection for breeding, the common English donkey might be gradually modified, and altered into a more symmetrical, stronger, and speedier animal than he is at present. The Spaniards breed both asses and mules of a larger sort than is done anywhere else. Their asses are particularly fine, and some specimens, which have been brought over here as 'curiosities,' have been known to excite the profoundest amazement of London donkey-owners. Goldsmith said of them : ' I have seen one over fifteen hands high.' Our horses and sheep and cattle of every sort have been improved in size and quality, all but the poor donkey, which, from sheer neglect, has deteriorated, or remains, at the best, no better than ever he was.

I certainly see no reason why asses should not be as capable of improvement as our other domestic animals, why their size should not be increased, and why they should not be rendered greatly more valuable to those who generally employ them as a source of profit. For the sake of these, our less fortunate neighbours and friends, I would like to see more attention paid to the breeding and rearing of asses. They are certainly one of the most useful kinds of animals coming within the means of the poorer classes. The importance of donkeys to these is considerable, both for breeding and rearing, and for working purposes, all which may add not a little to their limited resources, their comfort, and means of earning a subsistence.

It may seem a trivial matter, but I think the

following suggestion is worthy of notice. If landlords could be induced to keep a good jack-ass for the use and benefit of their poor tenants and others, whether labourers, costermongers, itinerant dealers, or what not, some good might surely accrue. These might then put their she-asses to a sound and picked mate, and thus get better stock than they do as it is. Something of the sort would be surely felt as a boon by the many who commendably maintain themselves and families, in some measure, from the service and breeding of donkeys. These are people who, belonging to the poorest class generally, stand most in need of such assistance, and, I doubt not, would gladly and thankfully avail themselves of it.

What has made the improvement in hunters, carriage and shire horses, but a like expedient? Landlords have been permitting their tenantry the use of a suitable stallion, usually free of cost, and the beneficial result this has had upon the stock of the country has been very perceptible. Other kind-hearted and noble-minded men, like the Duke of Westminster, have gone even further than this in generosity, adding to the comfort of labourers on their estates by affording them means of keeping a cow. To such persons, and to such as would emulate them, I leave the suggestion I have just made.

By such methods we may give, 'hoping for no

reward,' and I trust my hint will be taken up and an example made. For, if that were to be once done, I have little doubt it would soon be followed by many others, and thus the object might be gained of effecting a sensible improvement in the breed of asses.

If, in the absence of more direct illustration, I be allowed to fall back upon romance, let me refer to the touching tale of Sancho Panza and his ass. The love he exhibited towards *Dapple* is not overdrawn, but true to nature. Tears flowed from his eyes when reminded of his neglect of the poor animal; and afterwards, for the sake of the donkey's comfort, he even left a monarchy. In real life, were the condition of our asses altered for the better, I think we might expect to see them better cared for and attended to, more humanely and properly treated. And then the spirit reflected in the conduct of Sancho Panza might be manifested more often and more generally.

CHAPTER XXVII.

THE STUD FARM.—THE LAND.

Choice of locality and site—Aspect—Rich pasture not essential—Varieties of soil; examples—Ineligible sites—Cold and damp to be avoided—The land on which the best horses have been bred—Level pastures preferable.

Management of the grass—Feeding off with cattle and sheep—Proper rotation—Size of the enclosures—White-thorn and black-thorn hedges—Superiority of live fences.

Shelter-groves around the paddocks—Selection of trees—How and when to plant—Size—Undergrowth—Species to be avoided—Hedges and railings—The wrong sort of post-and-rail, and the right—Illustrations of both—Shape of the paddocks—Trees within them.

No work on the breeding of horses would be complete, or could lay claim to thoroughness, without dealing, more or less fully, with the subject of the stud farm. I shall therefore now go on to treat of this, using my own experience chiefly, but also that of others, for the demonstration of the views I entertain respecting all that may be required to constitute a first-rate breeding establishment.

One of the first considerations claiming our attention will naturally be the questions of situation and aspect, as well as the adaptability of the ground and its surroundings for the specified purpose of rearing young stock. Proper fitness in these respects I consider an object of paramount importance to be aimed at, and one that must be well discussed; for, if from any cause whatever we fail to secure suitability in these respects, ruinous failure must inevitably be the result. There are, undoubtedly, many excellent sites obtainable, in every way suited to the purpose of breeding thoroughbred stock. And here let me say that, in this chapter and the next, I shall deal principally, if not exclusively, with thoroughbreds and their requirements.

The character of the soil on the stud farm should be light and dry, preferably of light, friable loam. It should be kept in good tilth or heart by an annual supply of well-made manure from the stable, or compost of the same mingled with road-scrapings and ashes, if they are plentiful and easy to obtain. It is evident that very rich pasture is not essential, for the reason that the mares and foals will be supplied with succulent food otherwise, and with the most wholesome and nutritious provender in sufficient quantities in the stable. Here the grass is merely of auxiliary service, and it is not necessary that it should contain the richest possible elements. Yet I admit that it is a very beneficial adjunct, probably quite as good an one, or better, for the mares and their young than would be the luxuriant grasses grown on better land, upon which cattle are fattened.

I should prefer the kind of soil I have mentioned, since, upon it, the mares and their offspring would have the advantage of good dry air and healthy exercise. But it is not to be denied that other soils may also give good results. Even sandy soil, some of it, is quite well adapted for the purpose. That kind which abounds in Kent might be instanced. It was on such soil in that county that the late Sir Joseph Hawley, once known as 'the lucky baronet,' and the still more fortunate Lord Falmouth, reared their superior blood-stock. Yorkshire, too, stands high in the estimation of breeders as a county well suited for the raising of thoroughbreds. There the soil is rich, in many districts, resting, as much of it does, on a calcareous subsoil, or on sandstone. Mr. Wreford bred his horses in Devonshire on rich alluvial soil yielding grass that might have fattened a prize ox, and possibly sometimes did. Very similar in quality was this to the pastures around Doncaster, where Lord George Bentinck and other breeders raised valuable stock, where *Surplice, Loadstone, Pretender,* and *St. Hubert* were reared. Again, we have Mr. Chaplin's place in Lincolnshire, and Mr. Cookson's, Neasham Hall, in Durham, both places whence excellent stock has come.

It would seem, then, that favourable sites for the breeding of good thoroughbreds are to be found in various parts of the country, and are by no means confined to any one locality. Nor, indeed, can we say that they are confined to certain soils, to the loam, limestone, or sandy formations, though I think that these are the best, and will always be found the healthiest and best fitted for the purpose.

On the other hand, I should say that strong, undrained clays and marls, as well as boggy or heather lands, are certainly much the worst. Horses will not thrive on such ineligible ground. Low-lying lands, too, situated near stagnant pools or sluggish streams, are decidedly not conducive to the rearing of good sound stock. It could hardly be expected by the most ignorant that they should be; for delicate and susceptible creatures, like thoroughbred mares and their young, are sure to experience ill effects from the constant damp, fog, and wet of such localities. They require and enjoy a bright, bracing atmosphere, as much as they suffer from a cold or humid one. These requirements of their tender constitutions must be carefully studied and provided for, if we would ensure successful breeding.

Such wet districts as there are in many parts of Lancashire certainly do not appear well suited for the raising of thoroughbreds. But there are exceptions to every rule, and, even here, a *Mendicant* was bred. Much better localities are those found in Yorkshire, or on the chalk and limestone of the south.

The largest thoroughbred horses are often raised

on the best lands of the north, in Yorkshire chiefly. Those bred in the south are generally not so large, but I think they are mostly better. *Pretender* and *Doncaster*, both winners of the Derby, were two of the largest yearlings of such good quality I remember to have seen, except *Wild Dayrell*, who was bred at Littlecote, in Hampshire, and *Cadland*, bred in the New Forest, also Derby winners. I think that green sand, or fairly good land resting on a subsoil of chalk or limestone, is what we should select for the stud farm. It is on such land that good medium-sized horses—which I esteem as the best—are more frequently bred than on the best well-drained clay, or on alluvial soils.

The surface of the paddocks should be flat and level, which I deem better than if they are even slightly undulating. It is usual to choose ground of such character in preference to hilly or sloping land, which at times may be slippery, and so dangerous to the young stock. Very hilly pasturage, particularly when much rain has fallen after a prolonged drought, or in frosty weather, may be a fruitful cause of accident. Of such a kind were the paddocks of that estimable baronet, Sir Lydston Newman, at Mamhead, in Devonshire. His stud was a failure; not that I attribute the fact to this cause alone, nor, indeed, to it in more than slight measure. Various other causes contributed to bring about his ill-success, notably neglect of the duties of supervision on the part of his stud-groom. But that is apart from the subject under consideration. I merely refer to the Mamhead paddocks as an example of the wrong sort.

At least once a year, in the autumn, and oftener if necessary, sheep and cattle should be turned into the paddocks to crop them down bare. In Yorkshire one may often see a dozen bullocks, a score or so of sheep, and two or three brood-mares, all feeding in one paddock at the same time. For my part I prefer to keep them separate, putting the mares together in one paddock, and shifting them to another to let the bullocks follow them, and the sheep after the cattle in the same way. This system sweetens the grass for the following spring. Mares will feed after bullocks or sheep the same year, whereas they will not after other horses. In the same way, bullocks and sheep will feed after each other; but bullocks will not follow bullocks, nor will sheep follow sheep. Perhaps, when it can be so arranged, it will be best to let the mares have the first of the grass, then put in bullocks, and finally feed off with sheep; and, when the last have eaten the grass down close, they can be removed, and the paddock may be laid up for the winter.

When the land has been selected, and the pasture laid down, if it be not grassed already, the whole should be divided into paddocks of not less size than eight or ten acres each, and the larger the better—in reason, of course. The enclosures at Hampton Court are, for the most part, only some four or five acres in extent, and they, with many other establishments that might be named, are far too small, in my opinion. In regard to this matter foreigners are greatly in advance of us, and it is to the superior size of their paddocks that I attribute their success among us, far more than I do to any other cause.

The paddocks should be enclosed and separated with thick quickset hedges of white-thorn, or of holly, which is evergreen, and forms a compact, good fence. I lay stress upon the *white*-thorn, because Admiral Rous, in his work on 'Breeding,' erroneously recommends *black*-thorn. This last is really the worst material for a fence, as it is apt to run spirally upwards, and is wanting in the spreading, lateral branches of the other, which give thickness and resist the pushing and forcing of the animals enclosed. Again, the paddocks should not be separated by wooden fences, partitions, or walls. These are usually colder than are paddocks surrounded by live fences. However strange this may appear, it is an undoubted fact, I believe. At any rate, I am certain that horses do better in hedged paddocks than in the others.

According to the space that can be spared for the purpose, a deep belt of evergreen or other trees, with thick underwood of hazel and holly, should be planted on the north, east, and west of each paddock, leaving the south side only open, so as to get as much sunshine as possible. In light chalk and sandy soils, spruce and Scotch fir grow the best of any kind of tree we can select, and are ornamental and evergreen. Beech and deciduous larch might be planted with the others at first to act as leaders. The latter grows fast, would soon be fit for thinning out, and would be found very useful for repairing rails as, in time, they decayed and needed replacing. For this purpose it is excellent, as the wood is as durable as any fir. Where the land is good and suitable, oak, ash, and elm may be added to the trees already mentioned, and would stand for timber when the quicker-growing sorts were cut down or decaying. Privet, holly, and hazel would still serve as underwood, forming a compact barrier against the rough, cold winds of March and other inclement seasons. They should, of course, be cut as often as necessary, to keep them thick at the bottom, for strength and shelter.

The season for tree-planting begins in October, and extends until the end of February, or to March at the latest. During this time, so long as there is neither snow nor frost, the trees may be set in, the ground having been previously prepared for their reception by such trenching as the case requires. It is a mistake, and one often made, to plant them too deep. They should never be put in deeper than will just suffice to cover the roots. That this is sound practice may be seen from the fact that many of the finest forest-trees have been self-sown, the seeds having fallen on the surface of the ground and there taken root. The first year after being planted, and until they have taken root firmly, the young trees must be carefully looked to. After high winds or heavy rain this is particularly expedient, since many will then be found to require setting straight, or to have the ground about their roots well trodden down.

'To produce a large tree,' says an authority, 'plant a small one;' meaning that the young plant selected should be small of its kind when set, so as to develop to the best advantage. No doubt the advice is good, if one were planting for ornament. But in the present case we are chiefly concerned in forming belts for the sake of shelter, and the size the individual trees will attain to is not of so much importance. I should recommend that the young trees should be three to four feet high when put in. It is a suitable size for forming plantations to give good shelter in the shortest possible time. The trees should be set about four feet apart, and the interspaces should be filled in with hazel, holly, privet, and the like, for underwood.

Certain trees ought in no case to be found in the plantations about the paddocks. Savin, willow, barberry, lime, and yew are to be rigidly excluded from the neighbourhood of places where horses are kept, as being more or less poisonous. Many horses, and cattle likewise, have been hurt or killed by eating the foliage of these trees and shrubs. Leaves and branches might be torn off by the wind and carried into the paddocks, and so come within reach of the animals; while some people think they are more dangerous after having been detached from the tree and become withered. There are several varieties of yew, and perhaps all are not hurtful; but as some decidedly are so, it would be safest to avoid any.

Quickset hedges will be formed round the paddocks—of white-thorn, as has been mentioned—and should always be planted double; two rows being put in side by side, about six inches apart, and kept well trimmed. A mistake often made is that of putting the hedge too close to the trees of the plantation, whereas a clear space of six feet breadth should be left between the hedge and the grove. A similar error is that of putting the inside rail of the paddock too near the hedge, so that, while it is growing up, the horses are able to lean over and nibble the young shoots, thus spoiling the growth of the hedge. Here, again, a space of six feet should be left. Some ground will be undoubtedly lost by carrying out this advice; but I submit that the advantage gained outweighs that consideration.

Care should be taken that the inside post-and-rail fence surrounding the paddocks should present a uniformly smooth surface. This is seldom properly looked to, the rails being usually mortised

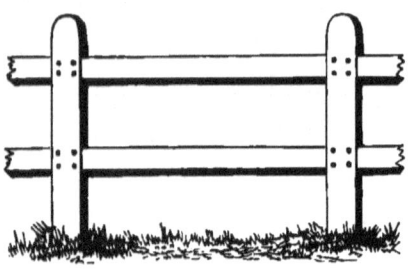

Fig. 1.

into the posts (Figs. 1 and 2), and so leaving an edge projecting several inches. This is enough to ruin any horse coming into contact with it, as may often occur when they are galloping round the

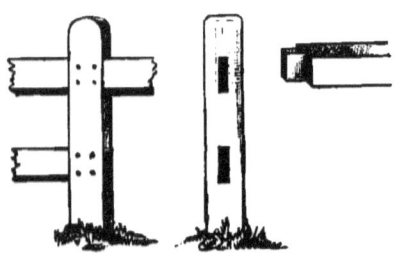

Fig. 2.

paddock. If the point of the shoulder catch against the edge of a post in this way, it may easily be injured past recovery. The best sort of fence is made by nailing the rail upon the post, and flush with it (Figs. 3 and 4). This kind has a neater appearance, and is cheaper, besides affording very much less chance of such accidents as I have alluded to.

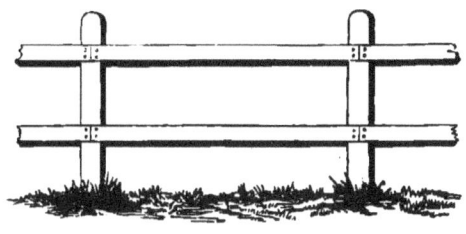

Fig. 3.

The shape of the paddocks should be oval or rounded, that is, there should be no sharp corners. This, though not perhaps an absolute essential, is desirable, at any rate. When laying out the shelter-

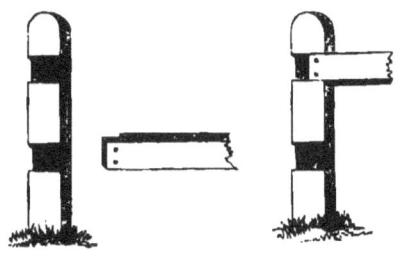

Fig. 4.

shrubbery, it will be easy enough to make thickets wherever corners would otherwise occur, and so to follow out the plan I have suggested. The object of this is to prevent the yearlings, when taking their gallops, from being sharply pulled up by an obstruction they may have failed to observe in their approach to it. Young horses are heedless as children, and might injure themselves by running into an abrupt corner, whereas they would gallop along a gradually rounding fence without risk.

Within the paddocks, here and there, may be dotted a nice specimen of oak, beech, or chestnut. The spreading boughs of such trees will afford shelter to the animals from sun or rain, and help to protect them from insect tormentors; besides which, they will diversify the landscape, and add beauty to it. At the same time, it is needful to remember that there must on no account be too many trees in a paddock, nor must they be close together. A very few, large and umbrageous, scattered over the expanse, will suffice. Anything more than that would be positively dangerous. Foals and yearlings will play and gallop in every conceivable direction, and might easily get injured by running against trees, if there were too many of them, or if they were too closely set. For, in the latter case, trying to avoid one, a rush of their companions might force them against another. The larger the paddock, maybe, the better, and the less reason, consequently, is there for fearing that the youngsters may hurt themselves by running against obstacles, whether single trees or fences. But however large the enclosure may be, it is better to have only a few trees in it, and those wide apart.

CHAPTER XXVIII.

THE STUD FARM.—THE BUILDINGS.

The master's house; central position—The men's cottages—The stables—Partitioning of the boxes—Cecil on the south aspect—My view of his theory—The roof—Collection of rain-water—Tanks and troughs—Shooting and guttering—Materials of the walls—Arrangements of interior—The manger and racks—Provision against accidents—Drains—Ventilation—Flooring—Doors.

The central yard—The various boxes—The roofing and the loft—Storage of hay, straw, and corn—Sleeping-room for stable-boys—This a necessary precaution—Boxes especially designed for stallions—Various materials—Cost of carriage—The furze hovel; how made—General arrangement of the stallions' boxes — Enclosed yards for exercise — Concluding remarks.

HAVING described the land suitable for the stud farm, its situation, aspect, and general treatment, it now becomes necessary that I should say a few words as regards the buildings necessary to complete a breeding establishment. There are various particulars in which my own experience, or my observation of that of others, may suggest some useful hints as to the erection, arrangement, and maintenance of the stables and their appurtenances.

The residence of the master, or overlooker of the stud-farm, should be situated as nearly in the centre of it as circumstances allow, and adjacent to the stables containing the brood-mares, foals, and yearlings. He will then be able to visit the various compartments without loss of time, and watch the animals more frequently and carefully than he could do if he lived at a greater distance from their stables and paddocks. So, too, their requirements will be seen and attended to more thoroughly; for, with race-horses, as with any other kind of cattle, the superintendent's attention has a marvellously beneficial effect, as has been not unwisely set forth in the old proverb, 'The master's eye makes the horse grow fat.'

The houses for the stud-grooms, stable-helpers, and others, should be placed as near to the stables as is conveniently practicable. They need not be an unsightly feature of the place, for they may be built in some spot out of the ordinary view of visitors who may wish to traverse the paddocks and inspect the stud, and they can be hidden from observation by trees. If expense is not so much an object, these habitations of the men may, of course, be built in ornamental style, so as to form an interesting feature of the place. Such matters must be left to the individual taste of those most concerned, not being of further consequence to our subject.

THE STUD FARM.—THE BUILDINGS. 379

The hovels, stables, or boxes, whichever one prefers to call the buildings standing by themselves, should be erected with four compartments, each with an opening leading into one of as many paddocks; and if they can be so constructed that all the doors face the south, so much the better. It would be the best aspect, undoubtedly, but I do not affirm that it is absolutely necessary. Round these buildings should extend a wooden fence or wall, making a small inner enclosure. Within this the yearlings, and even the mares too, may get some air and exercise at times when it is not safe to let them into the paddocks, as during inclement weather, hard frost, and snow. The erection of four boxes under one roof will prove economical, as the expense entailed will be not much more than half that of building detached ones, each being large enough for the accommodation of fifteen or twenty foals and yearlings, or a fewer number of mares.

I will here digress for a moment to refer to an experience of Cecil's, which is, I think, somewhat singular. 'I occupied,' he says, 'two paddocks in Shropshire, the soil of which was a sandy loam of fair quality, and although they were not far apart, the difference in the condition of the mares and foals occupying them was most apparent. Of course, they were all fed alike, and when the inmates were exchanged from one to the other paddock, a corresponding alteration in the appearance of the animals became very shortly perceptible. The paddock in which they did not thrive was enclosed by a brick wall, and, to the eye, appeared to possess every requisite—in fact, it seemed preferable to the other. It was screened from the northern aspect by rising ground and the house, and sheltered also by trees, which, far from being beneficial, were the reverse, by rendering the atmosphere damp—the ostensible cause, no doubt, of the unhealthiness of the situation. The hovels opened to the north, which I consider to have been another cause of evil. The other paddock, which I found to be so much superior, was more open; it contained three hovels placed in a row facing the south, an aspect which should always be selected for such buildings. Both paddocks were drained.'

But I never saw a race-horse, nor indeed any other, afflicted from the cause assigned by Cecil. Most stables are built either three-sided, or quadrangular, and may open or face in any direction. Still, I never saw nor heard of any horse which happened to stand in a stable opening to the east, north, or west, that was more frequently ill than others standing in stables opening to the south. There does not seem to be sufficient ground for being precise about this matter of aspect, and too much importance should not be attached to such remarks as those of Cecil. I should certainly prefer a stable to be so built that it should face and open to the south, but I would not make a *sine quâ non* of it. I should say it was a theory akin to that some people hold about the position of their beds, which they think should be set in the direction of the earth's magnetic currents, so as to ensure health to the persons who sleep on them.

But to return to the stable-buildings. The roof should be of slate or tiles, but with thatch underneath; or the rafters may be ceiled, so as to keep the stable moderately warm in winter, and comparatively cool in summer. Shoots ought to be put up to collect and carry off rain-water to a tank, which should be dug near the buildings for the purpose of storing the rain-water. This tank should be floored, walled, and arched over with brick-work, lined with cement, and should be of large enough size to contain water sufficient to supply the number of animals kept for at least two months, in case of a long drought. The water should be pumped up from the tank as required, into a trough of stone or iron, but *not* into a leaden one. This may be filled twice a day, being refilled after the animals have finished drinking.

The provision of a tank for rain-water, and adequate arrangements for supplying it, is a necessity far too often overlooked and neglected. If attended to, the animals are assured a proper supply of the best water, while much unnecessary expense is spared in fetching water to them, it may be for long distances, during dry seasons.

The roof of the buildings may be made entirely of thatch, which is both warm and economical. But, whatever the kind of roofing used, a proper provision for carrying off water is, under all circumstances, most advisable. Cast-iron shooting is the best, being more durable, and, in the long-run, cheaper than any other. The 'Half Round' is the most serviceable, but the 'O. G.' has the neater appearance. Zinc should not be employed, as gutters and pipes made of it are sure to warp under atmospheric influences, and constantly need repair. Wood, too, is sure to warp and crack in dry weather, so that the same objection applies to it. Guttering under the eaves has the further advantage of preventing water from dripping on the animals as they enter or leave the stable.

The outside walls of the buildings may be constructed of brick, stone, concrete, or mud, any of which materials will answer the purpose admirably. But whichever of these may be employed for the exterior walls, the interior lining should always be of one kind. Inside, the stable must be boarded to the height of four or five feet from the ground with half-inch boards of oak or elm—hard woods which few horses can hurt, or be hurt by, with gnawing. Some might prefer to employ a lining of deal, as being less expensive and more easily worked. In that case all corners, edges, and projections should be covered with a neat, well-fitted casing of zinc or hoop-iron.

Now, as to mangers. I think those of cast-iron are preferable to any others made, and probably the cheapest in the end. They should run the whole length of the box, and should be boarded from the outer top edge or bar down to the ground. This is a very valuable precaution, and ought to be always adopted. It prevents the possibility of animals getting their heads below the manger, rising suddenly or starting in the dark, and so injuring themselves. Accidents often occur—much more frequently than most people would suppose—in this way, from an outstanding, unprotected manger. It is a good plan, too, to have a cover for the top of the manger; a strong lid to shut down upon it, and prevent the risk of horses hurting themselves by kicking into it, and so entangling their legs. The lid may be made to open outwards and fasten to the front of the manger, at feeding-times, which would not be unsightly, and would be of great utility.

Above the manger, at distances of six feet apart, there should be iron rings attached to staples passing through the wall and secured by nuts outside. To these rings such of the horses as require control may be tied up whilst feeding. Greedy animals are thus prevented from disturbing the rest, or from depriving them of a full share of corn, of which all should have as much as they will eat. The rings, too, will be useful for securing the horses to whilst they are being handled and dressed. The manger itself should be so formed that the bottom be wider than the top, so that the outer top edge or bar shall protrude an inch or so inwards. The object of this is to prevent voracious feeders from pushing their corn out whilst feeding, and so wasting much of it, a fault which a good many are prone to. The same thing has the additional advantage of preventing the horses from hurting their knees against the under-boarding, which, if sloped inwards from the top of the manger to the ground, they might easily chance to do.

It is a bad plan to place the racks for hay above the heads of the animals. When so arranged there is the constant risk of seeds and small particles falling into the horses' eyes. Inflammation may be thus set up in those delicate organs, which, if not attended to, would be likely to result in blindness. A better system is to put the hay into the manger. Small iron bars fixed across the manger at intervals will serve to keep the hay in its place, and prevent horses from pushing it out and wasting it underfoot.

Drains are unnecessary in every kind of stable, and are more out of place in a breeding-hovel than in houses in which horses are more closely confined.

THE STUD FARM.—THE BUILDINGS.

Ventilation should be attended to, however. There should be, in every box, two openings close to the ground, and they should be so placed or guarded that they may not be the source of draughts. Another opening in the roof should do duty as a vent-hole for the escape of foul air.

All stables should be paved with flints or cobble-stones, rounded and firmly set in, and presenting a tolerably even surface. This kind of flooring gives the horses some foot-hold, so that they may rise easily, having something to rest the points of their hoofs against as they rise. It is, for this reason, a much better kind of pavement than one of bricks, wood, or asphalt, on which the horses' hoofs are more liable to slip.

The stable-doors should be at least five feet wide and eight feet high, and should be cut into two divisions, the top part being the lesser of the two. The door of the yearlings' stables should be of like proportions, but it need not be cut into two halves. There should also be windows spacious enough to afford ample light, say three feet square, or any other shape or dimensions, according to the size of the stable and to the number of horses it is intended to contain. There must also be air-holes for ventilation, on a similar plan to that I recommended in the former case.

The central yard of the stud farm should be square or rectangular in shape, facing towards the south, to which quarter it should lie open, the stable-buildings being on its north side, or on the north, west, and east sides. The boxes set apart for breeding purposes alone should never be less than twelve feet square; those intended for the yearlings, however, need not be so large; eight by ten feet, and ten feet high, being ample size for them. The same materials may be used in building as were recommended in the case of the paddock hovels, except that thatch is not a nice roof, and the internal arrangements may be similar. The roof should be of slate or tile, and there may be thatch underneath it if there is no loft, so as to ensure warmth. It is better so to build that the stables may be ceiled at a height of ten or twelve feet, with a floor above, forming a room or rooms for the storage of corn, hay, and straw. There will then be no necessity to under-thatch the tiles or slates of the roof itself. This loft should be boarded up to the eaves, so that the corn may be stored along the sides, in preference to placing it in the middle of the floor, which is naturally the weakest part. This precaution is not necessary, however, for the rooms given up to hay and straw, as they are not so weighty, and will require no moving after having once been packed in place for use as need requires. Oats, if stored in any quantity, will have to be moved in order to prevent heating, and to keep them sweet. They should be turned over once a fortnight, and should be winnowed once or twice just before being supplied to the horses. Something more may be added, too, in respect to the storage of provender. Both hay and corn keep better in the rick than in the loft, hence it is unwise to pack large quantities of either into the latter. Just enough of both should be kept there to supply what may be wanted for a few weeks, and avoid the possibility of running short in the stables. The horses, too, enjoy hay and corn more when they are fresh brought in, and derive more benefit from them then than they do after long storage in the loft. Though these are trifling details in themselves, they are nevertheless well worth careful consideration, and should be strictly followed.

If there is no house in the yard suitable for the purpose, a sleeping-room should be arranged in the end of the loft, and made a comfortable lodging for two or three of the helpers. These would then be always at hand, ready to assist any animal that might happen to get cast in the night, or to separate others that might, by some accident, have got together. We know that hounds are kept from rioting and injuring themselves or each other, in the night, by the timely assistance of a man who sleeps near their kennels. In the same way it is wise to take similar precautions in the stable. For, if the difficulties alluded to should arise and not be set right at once, some animal might easily be ruined before help arrived in the morning.

It now remains to describe the boxes set apart for stallions. Like most other erections, these may be constructed of almost any kind of building-material. If stone or brick can be had in the vicinity, either may be used. If the locality admits of it, chalk, or, as it is termed in some places, mud, can be employed for the walls. This material answers the purpose well; stables built of it are warm, dry, and comparatively cheap, while being sufficiently durable. In some parts of Hampshire and Devonshire, as also elsewhere, this kind of building-stuff is available, and answers well. Concrete is a similar class of material, which is found to be inexpensive and lasting, but, of course, only the first where gravel in sufficient quantity is obtainable at or close to the spot.

The great object in selecting the material to be used must be to avoid excessive expenditure. Carriage of heavy material for any considerable distance will prove more costly than many would be apt to believe; and therefore it is desirable to select that sort of stuff which is nearest at hand. I learnt this to my cost, and in a manner of which I have a very vivid recollection. For, when I was building my clock-tower at Alvediston, for which stone was employed, I found that the expense of carrying and preparing the stone came to double the actual value of it.

There is also a structure entirely of wood, roof and all, which is cheap, but not otherwise to be recommended, chiefly because the temperature of stables so constructed is apt to vary too much. I have found these wooden boxes generally too hot in summer and too chilly in winter, and therefore pronounce against them.

The last description of edifice I need refer to particularly is the furze hovel. To construct this, upright poles are driven into the ground and furze-branches are interlaced upon and between them. If the furze is tightly compressed and thickly filled in, a substantial wall is formed; and, when the sides are neatly trimmed and the roof properly thatched, the hovel is by no means unsightly in appearance. Moreover, it makes a warm and comfortable stable; and, wherever furze or gorse can be plentifully obtained, is decidedly the cheapest of any kind that will sufficiently serve the purpose.

Inside, however, the furze hovel should be boarded to a somewhat greater height than is done in other cases. Idle horses are too apt to amuse themselves by drawing out the furze with their teeth, bit by bit at a time, until at last they will make gaps and injure the structure. But a lining of boards will effectively prevent that, besides giving a more comfortable aspect to the interior than it otherwise would possess. The outside of the hovel may be protected from the same risk of injury by putting a railing round it, at a sufficient distance to prevent horses from reaching the wall.

The stallion-boxes ought to be larger, in all cases, than those of the mares and foals, yet they need not be so spacious as to make them uncomfortably chilly in cold weather. They should be situated as far as possible from where the mares are kept, for obvious reasons. They should be provided with all arrangements to ensure plenty of air and light. The doors, of similar size to those already spoken of, should be divided transversely in the middle, or rather about a foot above the middle, like those of the mares' boxes, so that, at certain times, the occupants may be shown individuals of the opposite sex. Each box should open into an enclosed yard, separated from other yards by a partition of brick or wood ten or twelve feet high. In these yards the stallions may be let out for exercise at the same time without seeing or being molested by each other, and here also may be performed the services for which they are kept. Each yard should be twenty or thirty feet square, at least, and longer than that if room can be given. Here the stallion may be allowed to take exercise whenever he likes to do so, on the turf or bare earth, as the case may be, enclosed in the yard. This will supplement the two or three hours' walking exercise which all stallions require but very few get, when they are taken out into the paddocks, or anywhere else out of sight of the mares and other horses. The stallion needs freedom to move about without restraint, and he enjoys his monotonous stroll round and round his private yard in fine weather. Only when the weather is inclement should he be deprived of this pleasure and confined within the limits of his box.

I believe that what I have now said on the subject of the buildings necessary for a breeding establishment will be sufficient. I am not an architect, and it is not my province to say more on the subject than to point out some features which I consider advisable. Individual taste must always have a good deal to do with the matter; but there are essential points which cannot be excluded from any design, and some of these I have endeavoured to indicate with such minuteness as was practicable in the present work. Breeders, or would-be breeders, whose means are ample, can perhaps afford to pass over much that I have said. Others, to whom economy in every respect may be of consequence, will possibly glean a hint or two from the foregoing remarks.

CHAPTER XXIX.

SOME ERRORS AND FALLACIES.

How able writers fall into error—'Stonehenge' on training—His mistake exposed—Buffon on inter-breeding of goats and sheep—His statement fallacious—Practical experience the only safe guide—Testimony of certain flockmasters—My own observations—The question answered in the negative—Why I have gone into the subject—The thoroughly practical man always the best authority—Obsolete works on breeding and training are misleading guides—Knowledge is progressive—A mistake of Goldsmith's confuted.

'Stonehenge's' opinion on the distinctions of brain in differently bred animals — This shown to be of no practical service—A case in point—Does blood indicate quality ?—Is there variation in the size and number of the bloodvessels ?—These views discussed and shown to be erroneous—Does early development induce early decay of horses ?—Argument and examples—I show this to be a fallacy—Proofs that our present system is a good one.

AUTHORS are prone to fall into error, and this may happen to even the most able and skilful, to those whose opinions are received by the many as authoritative, and to those whose general or particular knowledge cannot be lightly called in question. Not an infrequent occasion of mistake is that when a writer takes upon himself to discuss some subject with which he is not fully conversant, when, with the best intentions in the world, but simply not understanding what he is writing about, he may put on paper statements of an absurdly erroneous kind. Something of this sort must be the case when a writer on horses gravely represents a trainer as treating the animals under his charge in a manner quite contrary to that in which he should treat them, when he represents him as following a practice which any trainer would be careful to avoid, and so, by his own error, appears to recommend something which is really to be condemned.

'Stonehenge,' in his otherwise excellent and useful work on Training, has blundered in this way. The book contains a passage in which we are informed that, after a fortnight's steady work, a horse should be sweated four miles with two hoods and two or three cloths on, and another round his neck in some instances (a breast-sweater being meant, probably), and should be galloped at half-speed, and then much faster for the next two miles.

This is a capital error, arising from a want of practical knowledge on the part of the writer as to the ordinary methods of preparing a horse for a race. I should think that if a two-year-old or a three-year-old were to be treated in the manner just described, it would ruin or kill him, if he were in the backward state of preparation referred to and if the weather were hot. There is no need to say, therefore, that such a practice must be avoided. Of course, if the horse were a three-year-old, or older, and in a more forward condition, he might be sweated as 'Stonehenge' advises, though wearing only about half the clothing mentioned, without, perhaps, any injury resulting from it. But sweating has long ago been abandoned as part of the system of training, and therefore nothing more need be said with regard to it. I have only alluded to the passage in question for the sake of exposing the error, in order to prevent anyone from being misled by it. What follows will be introduced with a similar object in view.

One cannot always implicitly rely upon the information collected and credited by even the best and greatest authorities, men of extensive knowledge and of undoubted integrity and ability. Even these may trip occasionally on some point or other; and, despite what they pronounce to be fact, we must now and then exercise our own judgment and common-sense on matters with which we have practical acquaintance. To show that it is so, I will here give an illustration which, though apart from my main topic, may be very well held to bear some relation to it.

I find it stated by Buffon that 'the buck-goat is found to produce with the ewe an animal that, in two or three generations, returns to the sheep, and

SOME ERRORS AND FALLACIES.

seems to retain no marks of its ancient progenitors.' Now, here is a remarkable statement for such an authority to put forward. When naturalists are treating of points on which they have, personally, no practical knowledge, they should be careful not to promulgate mere theories as accepted facts. Moreover, they ought to refrain from advancing corroborative reasoning in support of their belief in such theories until the absolute truth of them shall have been established. For what is it but folly to pile up arguments in favour of something alleged, when that something proves to exist only in imagination?

The statement that the goat, being of the same order as the sheep, will breed with it, and that the produce is not sterile, is a grave mistake. There never was a greater fallacy put on paper. It is all very well to give scientific reasons in support of the assertion, to go into minute technical details which seem to prove the possibility of it, and then to lay it down as a fact that sheep and goats will intermix and their mutual progeny go on breeding. The argument is so convincing, that anyone might be pardoned for accepting the assertion as being true without further question. But is it so? No. Plausible as is the reasoning advanced to show why and how the thing might be and could be, what is the worth of it all when we find that the thing is not, and never has been, a fact?

Here it is, in this and similar cases, that the

kind of knowledge possessed by thoroughly practical men comes in. Only such can really pronounce finally in the matter. The practical man can say to it, '*I know*'—though perhaps, from want of learning, he may not be able to meet argument with argument, or dispute all the details set forth on the other side. In this instance of the goat and sheep, he may be quite unable to enter into any discussion of the anatomical and physiological identities of the two animals, which would make, in the scientist's opinion, fertile union between them possible and practicable. All he can say is that he knows it is not so; that the thing has been tried over and over again; and that goat and sheep together never have produced offspring at all, to anyone's knowledge; which, of course, puts the remainder of the assertion quite out of the discussion. All that is certain is that the he-goat will jump and ride the ewe, which may have misled Buffon or his informants, they not being aware that the connection—if there actually was any—has never been found to have result.

It is not a pleasant task, at any time, to write with the object of disproving that which another has advanced; and one feels especial repugnance at doing so when the other happens to be a truly eminent and justly admired authority. Nevertheless the task must sometimes be undertaken, however disagreeable it be to do it. Whether modern naturalists have or have not paid any attention to the particular statement of Buffon's now under discussion, I do not know. Such works as I have access to do not mention it. But whether they did or not, I should not have dared to enter upon the subject without thoroughly and practically investigating it for myself. I have taken the greatest pains to learn the truth of the matter, and will now proceed to set forth my views upon it at some length. It will be found that my denial of the circumstance Buffon alleges is borne out by the evidence of two large flock-masters whom I have consulted —men of well-known talent in their business, and whose respectability is beyond doubt. Their opinions and mine coincide, and fully confirm the result of my own observations.

What Johnson said of Shakespeare—which I am about to quote—might equally be said of Buffon. Indeed, the passage might be applied, with perfect propriety, to a great many cases. These are Johnson's words: 'Nature gives no man knowledge, and when images are collected by study and experience, can only assist in combining or applying them. Shakespeare, however favoured by nature, could impart only what he had learned; and as he must increase his ideas, like other mortals, by gradual acquisition, he, like them, grew wiser as he grew older, could display life better as he knew it more, and instruct with more efficacy as he was himself more amply instructed.'

Since Buffon wrote, the world has moved on apace, and much that was regarded as true in his day we now know to be false. But to return to the question as to whether goats and sheep can inter-breed.

A friend of mine, who is a practical man, and a noted breeder of Hampshire Down sheep for breeding purposes, told me that, in all his experience, he never knew of a single case of offspring got by a goat out of a ewe; nor had he ever heard of anyone else who had known of such a thing. He is the more to be believed since he was accustomed to let a he-goat feed with his flock, and this he would never have done if there had been the slightest risk of spoiling any of his valuable stock thereby. If there had been any such risk, or if he had even feared there might be, he would certainly have taken precautions against it. Thus, supposing he believed, as many do, that it was healthy for the sheep to have a goat or two among them, he could easily have taken care that these should be she-goats. Then, if any intercourse were to take place, the ram would not be spoilt, as a ewe and her offspring would most certainly have been in the other case.

When I lived at Woodyates, I have often, whilst exercising my horses on the downs, observed a he-goat, which was usually in company with the stock-sheep feeding there, jump ewes. The latter, when on heat, seemed disposed to admit him as readily as they would a ram. But though such a circumstance was not at all an uncommon event, I never heard of any ewe becoming pregnant by the goat.

Mr. Hayter, formerly of West Woodyates, who was a close observer of Nature, and whose practical knowledge of animals was very extensive, corroborated the foregoing as follows. He said that he-goats had been kept among his ewes, of which he had a flock numbering seven hundred, for many years past, yet he had never known them to cross. His son, Mr. Edney Hayter, also informs me that he-goats had also run with his flock, which was as large as his father's, for some years since then. Yet he was positive that no cross had ever resulted from intercourse between the two kinds of animal. Once he remembered to have seen a 'shock' (that is, a misshapen or abnormally featured lamb), which had some goatish resemblance while young. But 'shocks' are now and then produced in all flocks; and it was his opinion that, in this case, the goat had nothing to do with getting it. Such accidental resemblances, he said, have often occurred in other flocks with which a goat had never mingled; and therefore they cannot be ascribed to cross-breeding between the two kinds of animal.

In further corroboration of this last statement comes the testimony of a gentleman I am acquainted with, who has had considerable experience with sheep in Australia. He tells me he remembers a case where a lamb was found, running beside a ewe in a large flock, which, to all appearance, was a kid. It grew into a skinny, long-legged yearling, but was then killed or lost in the bush. The shepherds on the run considered it to be a 'back-slip,' an accidental return to some long-distant ancestral type. At any rate, my informant is certain there was no goat within several hundred miles of the district, nor ever had been.

Perhaps we may find, in the above-recorded statements, some explanation of the error into which Buffon fell. That a he-goat will mount an amorous ewe is one item; that a ewe will occasionally cast an abnormal lamb, having a goat-like semblance while young, is another. Mis-report or very superficial observation might, from these data, become the origin of an hypothesis which an unpractical man may have received for fact.

However, if he-goats have been allowed to run with large flocks of ewes for many years, as I have shown has been the case, and if no authentic instance of a cross between them has been recorded, what becomes of the positive assertion that 'they breed well together'? We may assume from the evidence the exact contrary, that goat and sheep cannot breed together at all; or that if they ever have done so, it has been an extremely rare and exceptional event, and one that did not result in the origination of a cross-bred race. But, be the probabilities never so great, no one should adduce theoretical ideas as matters of fact. It is deception in its worst form to do so when the topic is such an one as the present. For it is easy to see how people may be misled by following statements brought forward as ascertained fact, which have really no basis of truth in them whatever. Theories ought to be stated as theories, with just so much argument in their favour as may be reasonable and square with known facts. Others then will not be misled, but may, if they choose, resort to experiment in order to substantiate one side of the argument or the other. Valuable discoveries and improvements have resulted in this way; whereas nothing but evil can come from positively asserting what is merely matter of conjecture.

I have said thus much on this topic, to show how desirable it is always to follow the opinions and practice of an authority who has himself particular and intimate acquaintance with his subject, whatever that may be, and who has formed his views from personal experience or observation. To follow a mere compiler of various items is not always safe. We have just seen how a really eminent scientist, exact and reliable when dealing with things of his own knowledge, may, when writing from information received at second-hand, be deceived, and in turn deceive others.

I think it may be taken for granted that a man

who has followed any particular calling for a number of years is more likely to be well informed on practical details relating to that calling than another whose knowledge has been obtained by hearsay or otherwise. The first may not be erudite, may not be able to set forth what he knows in the polysyllabic language of the learned; but what he does say may be received with assurance, as having been derived from his own experience. In all matters of a practical kind, therefore, I hold that it is better to go to the practical man for instruction than to the most brilliant scholar, if the latter has not had also the former's opportunities of gathering special information.

Now, in illustration of this contention, I can imagine that a Chancellor of the Exchequer—the most accomplished who ever held office—would, if he were unacquainted with the methods of the Betting Ring, cut a very sorry figure there as a possible commission agent. With all his magnificent talents for dealing with a nation's revenues, he would simply be at the mercy of any 'knight of the pencil' who could bamboozle, bewilder and mislead him at will. In fact, the Chancellor would be as much out of his element in the Ring as would be the ordinary bookmaker, say, on the judicial bench.

I have considered it necessary to discuss this topic of the errors some standard authorities have fallen into, in order that I might show reason for

SOME ERRORS AND FALLACIES.

an object I had in view. That object was to warn my readers against placing too much reliance upon old and obsolete works treating of the breeding of thoroughbreds and other horses. Most of such books contain a great deal that is now useless, if not worse; theories long since exploded, alleged facts now known to be fallacies, or advice as to methods and practices which have now been wisely given up. On a practical subject, such as breeding or training, only the most modern works should be consulted, and they should have emanated from persons qualified to write them by special knowledge of the subject.

Goldsmith appears to have fallen into an error of a similar kind to that of Buffon in regard to interbreeding between goats and sheep. The English naturalist alleges that the urus or auroch, and the bison tribe, are of the same species as our domestic cattle, and can breed with them. More modern authorities have demonstrated, as the result of many experiments, that this is not the case. The aurochs and bisons can *not* breed with our cattle. But there is no need to go into the subject. It is simply another mistake arising in the same way as the one I have personally inquired into and determined to be a fallacy.

'Stonehenge,' writing of the greyhound, says: 'The chief distinction between the high-bred animal and his inferior consists in the brain and nervous system. It is true that we know little or nothing of the ultimate composition of the various parts of this system, but we judge by its manifestation that there is a difference, though we are unable to detect it.'

All this may be very true; I do not dispute it in the least; but I should like to know of what use it can be to *me*, a plain practical breeder. If the one sensible difference between the high-bred animal, whether dog or horse, and his inferior, is only to be discovered by inspection of their respective brains, the fact will be of little service to me, for I must kill my animals before I can know what they are worth. It is evident *I* must depend upon external characteristics, which shall be visible and easily discernible in the living animal. Else all these discoveries, important as they may be from a scientific point of view, must fail to be of the slightest practical utility. Is this sort of thing another fallacy or not?

I maintain that science, to be useful, must be brought down to the level of ordinary comprehension. The condition of an animal's brain can afford me no guidance in breeding from or using that animal, unless I have indications—whilst the animal is still living and the brain, therefore, not open to examination—of all that it is important for me to know. That I may make my meaning plainer, I will cite a case which I think will illustrate my ideas on this subject.

SOME ERRORS AND FALLACIES.

A horse may have disease of the brain, and yet, in the early stages at least, there may be nothing to indicate the calamity, or, if there are symptoms of something being wrong, we may easily ascribe them to some other cause, perhaps thinking them merely some passing ailment of no consequence. I recollect that this happened in the case of *Elmsthorpe*. He was sold when a two-year-old for £3,000, and at that time was probably already suffering from an unsuspected disease which shortly afterwards killed him. When he died it was discovered that his brain had softened till it had become of the consistence of cream, but nothing had been noticed previously that indicated the real nature or precise seat of disease.

As we cannot see the brain, or other great internal organs of the body, during life, we must be provided with other means of knowing whether anything is wrong with them. External conditions, or symptoms of some obvious kind, are necessary for our guidance. It is all very well to find out what was the cause of disease after death, but that, in itself, will not help us during life, and so far, appears to me to be utterly useless for practical purposes. It is only a step towards acquiring the kind of knowledge that will be of service to us with living animals.

The late Mr. Walsh ('Stonehenge') also seems, inferentially, to be of opinion that we can discover the condition of an animal by making an examination of its blood. If that be so, why has not such a great discovery been widely promulgated and made generally applicable? According to this theory, he says: 'We have only to take a little blood from either animal to ascertain the breed, from its odour, and the qualities of the horse or dog from which it was taken.' He goes on further to say: 'In the high-bred horse, as well as in the high-bred dog, the skin is thin and delicate, and the superficial veins are more readily seen. But these vessels are also really more numerous and capable of containing more blood, so that, during the very severe struggles of a long-continued gallop, the heart and lungs are relieved from the overwhelming quantity of fluid which would otherwise be dangerous to the safety of the animal. Hence, the blood has been taken as a test of high breeding, and has been supposed to differ in the form and composition of its globules, whilst the fact is, the difference really lies in the vessels which contain it.'

Surely this must be a mistaken idea. I do not profess to be a deep student of comparative anatomy, nor to have more than a rudimentary knowledge of the vascular systems of animals; but I have always understood, and still believe, that one greyhound is like another, one horse like another, and one man like another, in all their anatomical structure. So I should say that the whole system of bloodvessels, from the heart and aorta through arteries, capillaries, veins, and back again, must be identical in any two animals of the same species, as to size, capacity, extent, and distribution. One sort of greyhound cannot have a greater or less number of bloodvessels than another, according to all I have ever known of the matter. Exactly the same principle must apply in the case of any two horses, or any two of whatever kind of animal you like.

The internal structure of the horse must be, and is, always the same; except, of course, in regard to accidents of birth, malformations, and abnormalities. But, on dissecting a high-bred horse, I should just as soon expect to find in him two hearts, or three kidneys, as to find that he had more bloodvessels than another. And, in dissecting a base-bred horse, I should as much expect to find him without any heart or kidneys at all, as to discover that he had a less supply of arteries, veins, and capillaries than his superior congener.

What, then, becomes of the far-fetched hypothesis I have alluded to? Clearly it cannot stand the test of practical demonstration, and must therefore be dismissed altogether as a rank fallacy.

Another circumstance, well known to pathologists, is that, if by accident or disease the circulation of blood becomes impeded in one part of the body, as by the severance of artery or vein; the neighbouring bloodvessels distend, and get an accession of strength, so that through them the blood may be supplied to the part injured. This is a provision of nature to compensate for injury and effect repair.

However, we may be sure that we cannot depend on knowing the good or bad qualities of a horse, or perhaps his state of health either, from the colour, odour, or taste of the blood, nor from the number and size of the bloodvessels either. Nor can we do so except by the study of those external signs, characteristics, and symptoms which convey their respective meaning to the eye of an expert.

By many authorities early maturity is held to result in early decay. Hence, these persons suppose, the fattening of yearlings to the extent it is now carried proves a fruitful source of disease; and the premature training and employment of race-horses leads, they think, equally to premature failure of power. Here, as in most things, the happy medium is generally the best course to pursue. *Eclipse, Highflyer,* and other horses that were not raced until they were five years old, have been brought forward as examples in support of the above theory. I rather think they demonstrate just the reverse; for *Eclipse* was but two years on the turf, whilst *Highflyer* and the rest did but little to deserve a reputation for extraordinary powers of endurance.

Let us look at the history of some horses of our own times, and I think we shall see that early training and use of a thoroughbred does *not* imply early failure of its powers. Lord Jersey's *Bay Middleton* was not raced till he was three years old, yet he ran but one year; and many other owners did not run their horses until that or an older age. Among more recent instances we hear of various breeders (Mr. J. J. Farquharson was one such) who seldom had their horses broken in until they were four, five, or six years old. Yet none of the animals so treated ever proved themselves more capable or enduring than horses which had been broken, trained, and raced at a much earlier age.

Many of the horses Mr. R. Ten Broeck brought over here from America were not, we are told, broken until over three or four years old. *Preakness* was never ridden until he was eight years old. Yet none of these remained longer on the turf than such of our own horses as were broken when yearlings and raced as two-year-olds. For, of these latter, many have continued racing for thirteen or fourteen consecutive years, if not more, running in long-distance races too. We know that *Lilian* and *Reindeer* were examples of this kind.

In many cases, where our horses are good at two and three years old, some are never better than at the latter age, though continuing healthy and sound.

They will keep on running and retain their form, but they will not improve upon it, and so are often put to the stud early. The high feeding and care with which horses are treated at the present day will partly account for this, as well as the early time of year at which mares foal.

As to longevity, I find that none of the horses which have spent their first four or five years in idleness have attained a greater age than those which commenced work earlier. The *Godolphin Arabian* and *Eclipse* died before reaching the ages which *Touchstone* and *Pocahontas* lived to see. It seems to me, then, that the evidence is all in favour of early breaking and high feeding, if sufficient exercise be added to the treatment from the outset.

If horses be made as fat as a Christmas ox, you can hardly expect that they will not be subject to disease of some sort; for excessive fat is itself a malady. In the natural state they are but little subject to ailments, and are never then found in a disgustingly obese condition.

The fact that no more horses die from influenza, catarrh, or bronchial affections than is the case, speaks well for our present systems of management after they leave the breeder's hands. They are now burdened with more fat than their ancestors were in former days, when they used to be kept running loose in the paddocks, and were put into training lean and spare. We feed them better now and work them earlier, and the results seem to prove that our present methods are good ones.

Experience has shown that the change from one extreme to the other was a bad mode to follow, and such systems have been wisely abandoned. In other ways, too, our horses are better provided for than used to be the case. Stables are kept cooler, and at a more equable temperature, while the art of the veterinary surgeon has so progressed that the lives of many horses are saved which would otherwise have been lost.

No doubt many yearlings—not all, happily—are over-fed and under-exercised before they come into the trainer's hands. Of such, not a few soon fall victims to disease, possibly to fatal disease. This is a fact that must have come under the observation of almost every trainer. But the maladies induced in yearlings by over-feeding and want of exercise are met with more and more understanding and talent on the part of our veterinary surgeons, whose knowledge of 'all the ills that flesh is heir to' has vastly progressed with the passing years. They know better now than they used to do how to fight incipient disease with proper and adequate remedial measures. Were it not so, the loss of young or comparatively young horses would be more serious than it is; though a great many more are still lost through ignorance and folly than most people suppose, or than need be the case.

CHAPTER XXX.

SOME FINAL SUGGESTIONS.

Too many stallions—Statistics of 1882—A recommendation—Startling figures—Mares also too numerous—Existing bad qualities—How to better them—Evils propagated—The French system—Bad temper—How to remedy defects—Export sales.

Statistics of 1884—Fees—Small proportion of good stallions—A warning to breeders—Mares that should be excluded from the 'Stud-Book'—Excellence not properly kept up—Injudicious foreign sales—The stallions serving in 1884—Inferiority of the majority.

Fifteen thoroughbred stallions to be selected annually—Twelve years at the stud—Proportion of mares to stallions—Modes of choice—How to get rid of weeds—Evils of cheap breeding—Two cases in point—Reform desirable and necessary—The probable result.

I REMARKED, in a former chapter, that the number of stallions kept in this country was far too great. There is no doubt that this is really the case. If half the stallions that are kept every year were cut as yearlings, great improvement in our breed would follow. Let us look at some figures and see what they teach.

Referring to 'The Racing Calendar,' or 'The

SOME FINAL SUGGESTIONS. 413

Stud-Book,' I find there were foaled, in 1882, of thoroughbred colts 920, and of thoroughbred fillies 996. In some years the colts have outnumbered the fillies, as in 1833-4, in both which years there were over a thousand of them. Taking one year with another, there is probably not much difference in the numbers of the two sexes respectively, and the average of each may be set down as about 900.

I would suggest that, of these 900 colts, one-third, or 300, should be castrated, and the like number probably may be disposed of abroad. This would leave us 300 for use, more than enough, as I will show. The average number of mares at the stud may be set down as being 3,300, taking one year with another; and for these 84 stallions would be enough. For, we may well allow 40 mares to a stallion, which would not be too many or overtax his strength. Add to these 84 stallions another 16, to supply deficiencies occasioned by death, disease, incapacity, or whatever cause, which would be giving ample margin, and this 100 would be sufficient for the specified purpose. We have then still 200 stallions left. Some of these might be remuneratively employed in breeding half-breds for hunters, carriage-horses, or, indeed, as I have previously mentioned, for nearly all light draught purposes. If all could not be profitably employed in this way, the excess might be added to the number exported, as there is generally a sufficient demand for them abroad.

Further pursuit of this interesting topic will reveal some startling figures and calculations, from which the only deduction possible is that a still more extensive use of the gelding-knife is required. Let us allow that each stallion remains twelve years at the stud, which is, I think, a reasonable average time to reckon upon. The yearly additions of 300 fresh horses would swell the total available for service to 3,600. These, giving the former allowance of 40 mares to each horse, would suffice, therefore, to serve 144,000 mares annually, thoroughbreds and others. And these, be it remarked, are only the *thoroughbred* stallions available for stud purposes. We have yet to take into account the probably much larger number of half-bred and other horses which are travelling the country everywhere between John o' Groat's and the Land's End, serving common-bred mares.

As things actually *are*, there are more than double as many thoroughbred stallions kept for use as are really wanted, many serving, no doubt, at an exceedingly low figure. In 1882 there were only 75 advertised in 'The Book Calendar.' I suppose the number was so small because even the infatuated owners of the large remaining majority did not think they were good enough to defray the cost of advertising them. Out of these 75

SOME FINAL SUGGESTIONS.

stallions, presumably the best, or held to be so, of all that were serving, 33, or nearly half of them, are not recorded as having got a winner that year; 7 got only one winner each, and 14 not more than five. Of other stallions besides those advertised in 'The Book Calendar,' there were 176 named as sires of winning horses, though, like the above, they were not all got in the same year. Ninety had one winner each, and 47 got two, or at the most three each; also there were twelve mares which had been covered by more than one stallion, and they each bred a winner. I see no account of the remainder, so I suppose none of their progeny were honoured as winners.

There are also undoubtedly far too many mares unfitted for breeding thoroughbred stock, which are yet kept for the purpose. The produce of these, after hundreds of pounds have been spent over them are only fit for inferior purposes, and drag out their wretched existences between the shafts of a London cab or butcher's cart. Yet useless mares are not so harmful as too many stallions. The latter might be estimated to get forty times as many foals as themselves annually; the mares would only breed one apiece at most, and so would not burthen the country with anything like so large a proportion of animals as useless as themselves for the chief purpose of their existence.

Of the stallions lately mentioned which did get

winners, several had curb, some were roarers, and no less than eleven were bad-tempered, many of their progeny turning out equally faulty. Others, again, were light of bone, upright in the pastern, twisted in the toes, calf-knee'd, and otherwise deficient in the qualities necessary to render them good and sound at the stud. Yet to such horses, with defects of the most obvious kind, people were found ignorant or foolish enough to send mares. Under such circumstances the ensuing failures were only what might have been expected. When racing, too, many of the said stallions went lame from various causes: splint, thoroughpin, curb, and spavin, contraction of the feet, or navicular disease. Others were undersized or overgrown, and in many ways totally unfitted for the stud.

I think that if gentlemen who breed their own horses would but have every moderate yearling cut, as well as all such as had any serious defect as to shape, and kept only such entire as were well formed, sound at all points, and of the very best strain, it would be an immense improvement. I think posterity would hail such innovators as benefactors to the breed of horses, and consequently to the country generally. I will go farther and say that if, of the geldings thus made, any turned out well, they would prove of more service for racing purposes, and be more valuable in such capacity than they would if they had been suffered to remain entire, and, after winning a few races, been sold for the stud. For geldings will certainly last much longer on the turf than either mares or stallions, and may consequently win a larger proportion of races.

It appears to me that bad-temper is a fault almost entirely overlooked by breeders. To this drawback I attach the very greatest importance, as I have explained in a former chapter. Out of the number of stallions just cited which got a winner, no less than eleven, as I have said, were bad-tempered. Now these, if they were lucky at the stud, and their subscription lists well filled up, might annually get four or five hundred horses with tempers as bad as their own, or worse, perhaps.

We may and do see the evil that so many country stallions are engaged in propagating, to the injury of our breed of race-horses, and heartily condemn it. Yet, say what we will and do what we may, it continues, as the number of weeds at the stud fully demonstrates. I believe nothing will put a stop to this until coercive measures are introduced, and Government inspectors appointed in every county, duly authorized to condemn all unfit horses to the gelding-knife, or to prohibit the use of them as stallions.

The idea just mentioned is no new one, but is, I conceive, one worth attention and advocacy. In France, before the owner of any stallion may permit his horse to serve mares, the horse must be passed by a Government inspector as free from hereditary disease, and be otherwise fully approved of. This plan ought to be strictly followed in England, if we wish to maintain our supremacy as breeders of the best horses in the world. It may be an interference with the liberty of the subject; but not more so than exists in many other ways. We have veterinary inspectors and magistrates invested with power to order glandered horses to be destroyed, and to stamp out other contagious diseases in cattle and swine. The same powerful means being used, why should not equally effective measures be taken to prevent stallions ineligible in a variety of ways from entailing imperfections on their progeny, and the same with mares?

'Tom, the brother of Jack Clarke,' says 'Nimrod,' 'after sweating a grey horse that belonged to Lord March, with whom he lived, whilst he was scraping or dressing him, was seized by the animal by the shoulder, lifted from the ground and carried two or three hundred yards, before the horse loosened his hold. *Old Forester*, a horse that belonged to Captain Vernon, all the while I remained at Newmarket, was obliged to be kept apart, and to live on grass, being confined to a close paddock. Except Tom Watson, a younger brother of John Watson, he would suffer no lad to come near him. In a match of four miles he seized his opponent by

SOME FINAL SUGGESTIONS.

the under-jaw and held him back, though he lost the race.' We also have in *Paradox* a modern savage, like his grey prototype, if the accounts we hear of him, as to his dislike of his attendants, are reliable.

In order to improve the size and formation generally of horses, we know that in olden times no stallion below a certain height was allowed to be turned out on the different commons to graze with other horses, should mares be among them. If this law was contravened, the owner would, for his negligence or wilful disregard of it, suffer the loss of his horse, a very arbitrary but salutary statute; yet no better method could possibly have been devised for enforcing the regulation. What was then found so beneficial would now be the same, perhaps to a greater extent, for the improvement of our present breed of horses; not for increasing their size, as that object has already been attained, but for remedying various known defects in their constitution, which it must, to every observer, be manifestly apparent that they stand in great need of. Castration in those days completely answered the purpose for which it was intended, but to it now we are obliged to add expatriation. Still, if we are to induce foreigners, our best customers, to continue purchasing our horses, we must produce stock that will be worth their attention, or they will be driven to France or

America in search of such horses as they could once obtain here. Our markets, too, will soon be glutted with a set of animals useful neither for racing purposes, the saddle, nor anything beyond common harness-work. The best way I see of ensuring profit would be to have at least two-thirds of our thoroughbred horses cut, and half of the other moiety used only for improving the breed of hunters, and other descriptions of riding-horses. In 1884 there were 90 stallions that were thought good enough by their owners to be really worth the expense of advertising in 'The Racing Calendar,' their covering fees ranging from 5 guineas to 150 guineas. Now, if the service of any horse in covering a mare be worth 150 or 250 guineas, at which last fee *Hermit* is covering, it must be apparent that 5 guineas or 2 guineas, a price that many horses serve half-bred mares at, cannot be a remunerative or fair price for the service of such another, as is, or ought to be, capable of getting good thoroughbred stock also. It shows too plainly that the services of such stallions are not required, for they can only be expected to breed a still more degenerate race than themselves, particularly when mated with mares of much the same class. Only through the extreme lowness of the price they cover at are any induced to use them.

If this argument holds good, and I think it will not be easy to rebut it, what can we reasonably expect from 300 still less carefully bred horses, possessed of greater defects in a variety of ways, and thus rendered absolutely unfit for propagating high class race-horses, the purpose their owners intend them for? It would be invidious to point out (in the list of stallions appended) what are the particular defects of each respectively. One may be badly shaped, another affected by hereditary or incurable disease, and others afflicted, in one way and another, so as to be certainly unsuccessful at the stud for getting good thoroughbred stock. Yet I would call attention to this important matter, in order to warn breeders. I would seriously caution them against the use of such horses as are known to be afflicted with curb, ring-bone, spavin, or thoroughpin; or that are roarers, or overgrown, flat-sided animals; or that are undersized, galloways, or ponies, such as we know to have been bred from horses of bad constitution. I would caution them, too, against unsound or in any way malformed animals, and, above all, against bad-tempered and savage horses. Nor should such as have proved themselves bad on the turf be bred from, except in rare cases where they were of very superior breed, good shape, and had successful progenitors on both sides.

Now, I think that not only one of these faults will be found in at least three-fourths of the stallions covering in 1884, but that in many will be found a complication of several of them. So that, perhaps, two-thirds of the list consists of horses quite undesirable as race-horse sires.

If the names of thoroughbred mares which have ever been covered by a half-bred horse, by a Turk, Barb, or Arabian, were to be rigidly excluded from 'The Stud-Book,' it would be the means of keeping the blood a little purer. This rests upon the fact, which I have proved in a previous chapter, that, after mares have been so covered, they will never breed so good a foal to any other horse as they might have done before. At any rate, the plan would have the effect of getting rid of some 60 or 70 mares that have been served by half-bred horses, or by Barbs, Turks, and Arabians, that year by year are breeding 40 or 50 foals with a stain in their blood, all of which ought to be wiped out of the lists of our thoroughbred horses.

Out of the 1,001 colts bred in 1884, we find no less than 336, or more than one-third, are thoroughbred horses kept to breed from. How many half-breds there may be besides—I imagine they are not a few—that are kept and bred from by English breeders, it is not easy to say. On what principle they are used I am at a loss to conceive also. If, as a rule, we only breed a few good horses, perhaps a dozen a year, how can we expect to perpetuate excellence by breeding from over 300 inferior animals, which to all intents and

SOME FINAL SUGGESTIONS. 423

purposes they must be, comparatively speaking? I am sure no flock-masters or breeders of prize cattle would pursue such a haphazard course and expect success. This assertion may seem a little extravagant, but it is founded on sound principle, I am sure, and if we kept only 15 stallions a year, since each of them would last at the stud, on an average, 12 years, we should have in the course of that time 180 of the best stallions that we could produce. And as one horse is enough for 40 mares, the 180 would be sufficient for 7,200 mares, or nearly double the number we have put to the horse annually for breeding thoroughbred stock, and this would leave abundance to fill vacancies caused by death or illness. The superabundant animals we breed yearly, amounting on an average to something like 321 stallions, except a few that are cut, may be put to improve any other descriptions of our horses, for which they may be most fitted, or for exportation, with the other 300 before mentioned.

It is the selling to foreigners of so many of our good stallions and mares that has forced us to breed from inferior stock, and has given them at times a seeming advantage over us. A glance at the number and quality of the horses we export annually will prove the fact. Among them were *Buccaneer* and *Musket*, also *The Bard*, sold to France at the enormous sum of £10,000. If we had only good fair stallions or mares to dispose of, it is probable we should still have plenty of customers. For though they could not obtain the best, they would purchase those next to the best, which at any rate would be better than most of their own, or than what they could purchase elsewhere, and so breed from them rather than be without the advantage of the blood altogether, on the principle that 'half a loaf is better than no bread.' This may be proved by the fact of their purchasing a roarer, coming from a roaring family, namely, *Prince Charlie*, that in shape and size was not calculated to get race-horses. I may say this much, as the opinion I entertained long ago, and, now he is dead, there can be no reason against my openly expressing it.

I give in an appendix the names, alphabetically arranged, of the whole 320 stallions which I find were covering in 1883-4, so that readers may judge for themselves of the respective purposes such animals are most fitted for, many being half-breds, Barbs, or Arabians, and other descriptions of foreign horses, and these not a few in number. I venture to think among them there are not many, comparatively speaking, known to fame, either on the turf or at the stud. By the method proposed we should, though slowly, be assuredly progressing in the right direction, which would ultimately culminate in success, not only for the improvement of our own race-horses, but also, it would be found, in

SOME FINAL SUGGESTIONS.

enhancing the value of every description of horses. So would our reputation as the breeders of the best horses in the world be maintained.

Now, let us, as before, assume the average number of mares at the stud to be 3,300. If we select from their produce 15 stallions annually, and I am sure there are not more, if so many, really good horses bred in each year, we should have 180 first-class horses at service. For, as before remarked, I consider 12 years the average duration of a stallion at the stud. The number of thoroughbred mares used for breeding thoroughbreds exclusively might also be reduced with advantage. But the 180 stallions, to each of which 40 mares might be allotted, would be capable of serving 7,200 mares in all; and that is more than double the number we have at the present time in use at the stud.

Of the remaining stallions bred annually, let us assume that 185 are kept entire, and not exported. These, in twelve years, would give us 2,220 thoroughbred horses, employed in multiplying half-bred hunters, hacks, carriage and draught horses. But then we must take into account half-bred stallions, and such as may be kept of other descriptions. These would be more numerous than the thoroughbreds, and would surely make the aggregate number of stallions 5,000 at least, sufficient for the service of quite 200,000 mares.

Of course, I can see there might be difficulty in selecting the 15 stallions of the year. But there are many ways in which this might be done, and I do not think the difficulty would be insurmountable. Suppose a score or two were found to be properly qualified, the 15 for use could be balloted for from among them. It would be made a Government undertaking, of course, so far as inspection went, and I have no doubt would prove remunerative to the country. The winning of certain races might be taken to render stallions eligible ; or the Jockey Club could find plenty of other means to the same end.

I would also suggest that all mares not served exclusively by some one or more of the privileged 180 stallions should, for that reason, be cut out from ' The Stud Book.' This would help us to get rid of many badly-bred weeds, such as are now at the stud, whose stock does not fetch half the money it cost to breed and rear it. Owners would no longer care to give a fee of £50 or £100 to have a mare served, unless she was a good one. Nor, as is now the case, would retired tradesmen, who are fond of the turf, deem it as easy to breed the winner of the Derby from any defective little mare they can pick up cheap, as to buy a spavined jade, and use her in their delivery-vans.

A check would in the same way be put upon those other wiseacres, who prefer to breed from a stallion that has been parted with cheaply on account of disease or defect which really renders him quite unfit to be a sire. Of course he may turn out another *Merrymaker*, and get innumerable *Jesters*, but that is not very likely. His owner sends his own mares to him, in the full expectation of realizing his most unlikely dream, and succeeds in persuading his more simple neighbour to follow his own enlightened example. It is only three or four years ago I saw a stallion knocked down at a public auction for 2 guineas, and he is actually sire of a horse that ran and won at Newmarket in 1886. I lately saw another instance of a really well-bred horse which never ran, having two forelegs that would scarcely enable him to walk, let alone gallop, being as straight as two broomsticks, and not much larger in circumference, whose owner had better have shot him than ever allowed him to cover one of his mares. But he thought differently, and put a few to him, and from the fact of his getting a winner out of a mare that had previously bred one or more, he was at once confirmed in his opinion that he had a second *Hermit*. Therefore he continued putting his mares to this horse, which so far has been, and is still likely to be, disastrous. Probably the two-guinea stallion above-mentioned will, in after-life, be much about the same, yet these two stallions have a number of mares, if not the whole stud belonging to this couple of geniuses, put to them year after year, besides other mares belonging to different owners.

Some such reforms as I have indicated seem more than desirable, even imperative. I cannot calmly contemplate breeding from imported horses or mares. Our own horses are already too much mixed with foreign blood, of whose breed, constitution, temper and soundness we absolutely know nothing, or how degenerate and unhealthy may have been the stock from which they came. How, in breeding from such animals, can we expect to do anything but to cram the country with a race of horses that absolutely may not be free from any one hereditary infirmity or constitutional defect, and in the course of a few years to find that the discovery of a sound and useful horse for any general or specific purpose will be a matter of notoriety? Even of our own horses and mares too many are not fitted for the purpose of breeding, and should only be used as long as they can be profitably employed in drudgery, for which only they are fitted; or be transported beyond the seas, where, in the absence of better, they may have admirers, who may find them of more service than we should here.

From breeding in such a haphazard sort of way, as many do, without even an elementary knowledge of the science, what can be expected but the fatal results I have described? With a less number of mares and stallions, I have little doubt, if they are properly mated, it would be just as easy to breed more good runners than bad, as it is now for us to have a superabundance of animals that are really not worth rearing. The dams and sires of such animals should be put to another purpose, such as to breed light riding or phaeton horses, or other draught-horses, for which they are more suited than for breeding race-horses, or they should not be allowed at all to propagate a degenerate race worse than themselves. If a third of our mares and a like number of our stallions were excluded from the pages of 'The Stud Book,' and no longer looked upon as thoroughbreds, we should probably have better horses, and more of them, than we have now, for we may continually see bad mares going to good stallions, and good mares going to bad stallions, and the inevitable result is failure in both cases. Keep only such mares as are fitted for breeding purposes, and put them to such stallions as are likely to suit them, and thus a better race will be reared, and the highest aims of the breeder be realized.

APPENDIX I.

THE ALVEDISTON STUD IN 1873.

(1) STALLIONS.

1. CAMERINO, brown horse, by Stockwell out of Sylphine, by Touchstone out of Mountain Sylph, by Belshazzar out of Stays, by Whalebone. He is the sire of Quick Step, George Osbaldeston, The Prior, Herminie, and other winners.

2. MAN-AT-ARMS, bay horse, by Kingston out of Paradigm (dam of Lord Lyon and Achievement). He is the sire of Savoir Faire, Sentry, and other winners.

3. PROMISED LAND, brown horse, by Jericho out of Glee, by Touchstone out of Harmony, by Reveller. He is the sire of Cast-off, Juanita, Rather High Colt, and other winners.

(2) BROOD MARES AND FOALS.

4. AERIAL LADY, b., 7 yrs. old, by Wild Dayrell out of Odine, by Fitz-Gladiator, her dam Pauline (Fille de l'Air's dam), with Filly by Man-at-Arms (foaled May 3), covered by him again.

5. ARABY'S DAUGHTER, br. (dam of Oxonian), 19 yrs. old, by Flatcatcher, her dam Macremma, by Sultan out of Dulcinea, by Cervantes, covered by Y. Trumpeter.

5*. Ch. COLT, by Oxford out of Araby's Daughter (own brother to Oxonian). Foaled March 30.

6. AUSTRALIA, b., 8 yrs. old, by Vedette out of Goldfinder's dam, by Liverpool, out of Ninny, by Bedlamite, with Filly by Camerino (foaled February 2), and covered by him again.

7. BERENICE, b., 5 yrs. old, by Trumpeter out of Lady Williams, by Knight of Kars, her dam Marpesia, by Bay Middleton out of Amazon, by Touchstone, with Filly by Camerino (foaled February 11), and covered by him again.

8. BIGLY, ch., 11 yrs. old (dam of Savoir Faire), by Neasham, her dam by Pantaloon out of Banter (Touchstone's dam), covered by Camerino.

9. BIJOU, ch., 4 yrs. old, by Trumpeter out of Regalia, by Stockwell out of The Gem, by Touchstone, her dam Biddy, by Bran out of Idalia, by Peruvian, covered by Y. Trumpeter.

10. BURGAS, bl., 7 yrs. old, by Vedette out of Varna, by Venison out of Odessa, by Sultan out of sis. to Cobweb, by Phantom, with Filly by Promised Land (foaled January 27), covered by Camerino.

11. CASTANETTE, b., 17 yrs. old, by Pelion, her dam Concertina, by Actæon out of Brocade, by Whalebone, covered by Y. Trumpeter.

12. CHIC, br., 8 yrs. old, by Stockwell out of Sprightliness, by Touchstone, her dam Columbine, by Plenipotentiary out of Corumba, by Filho da Puta, covered by Y. Trumpeter.

13. COCHINEAL, br., 14 yrs. old, by Sweetmeat, her dam Biddy, by Bran out of Idalia, by Peruvian, covered by Camerino.

14. CRACOVIENNE, ch., 7 yrs. old, by Trumpeter out of Cachuca, by Voltigeur, her dam Ayacanora, by Birdcatcher out of Pocahontas (Stockwell's dam), by Glencoe, covered by Camerino.

14*. COLT by Camerino out of Cracovienne. Foaled March 18.

15. CROSSFIRE, b., 11 yrs. old, by Vedette out of Cross-

lanes, by Slane, her dam Diversion, by Defence out of Folly, by Middleton, covered by Camerino.

15*. COLT by Camerino out of Crossfire. Foaled February 23.

16. DAME SCHOOL, ch., 4 yrs. old, by Stockwell out of Preceptress, by Chatham (her dam Oxonian's dam), by Laurel—Flight, by Velocipede, covered by Man-at-Arms.

17. DECOLLETÉE, ch., 6 yrs. old, by Marsyas out of Gossamer (dam of Miss Foot, Bel Giorno, etc.), by Birdcatcher out of Cast Steel, by Whisker, with Filly by Man-at-Arms (foaled March 25), covered by Camerino.

18. DELILAH, br., 19 yrs. old, by Touchstone, her dam Plot, by Pantaloon out of Decoy, by Filho da Puta, covered by Camerino.

18*. Ch. COLT by Camerino out of Delilah. Foaled January 20.

19. DRIED FRUIT, b., 4 yrs. old, by Stockwell out of Fravola, by Orlando, her dam Apricot, by Sir Hercules out of Preserve, by Emilius, with colt by Man-at-Arms (foaled May 5), and covered by him again.

20. FIRST LADY, b., 8 yrs. old, by St. Albans out of Lady Patroness, by Orlando out of Lady Palmerston, by Melbourne, her dam by Pantaloon, covered by Y. Trumpeter.

21. FREA, ch., 11 yrs. old, by Ethelbert out of Braemar, by Greatheart, her dam Highland Fling, by Venison out of Reel, by Camel, covered by Man-at-Arms.

22. HOPELESS, br., 8 yrs. old, by Vedette out of Hope, by Touchstone out of Miss Letty, by Priam, with Filly by Man-at-Arms (foaled April 7), covered by Camerino.

23. JUSTITIA, b., 8 yrs. old, by Macaroni out of Independencia, by Y. Melbourne out of Flirt (foaled in 1855), by Orlando out of Elopement, by Velocipede out of Scandal, by Selim, covered by Camerino.

24. LARGESSE, b., 16 yrs. old, by Pyrrhus the First (winner of the Derby) out of Mendicant (winner of the Oaks), by Touchstone out of Lady More Carew, by Tramp, with

Filly by Camerino (foaled March 8), covered by Y. Trumpeter.

25. LA TRAVIATA, b., 17 yrs. old (dam of Violetta), by The Flying Dutchman out of Boarding-school Miss, by Plenipotentiary out of Marpessa, by Muley, with Filly by Camerino (foaled April 25), covered by him again.

26. LUCRETIA, br., 6 yrs. old, by Voltigeur out of Village Maid, by Stockwell out of Minx (sis. to Melbourne), with Filly by Promised Land (foaled May 4), covered by Camerino.

27. MAID OF ATHOL, ch., 6 yrs. old (sis. to Jock of Oran), by Blair Athol out of Tunstall Maid, by Touchstone, her dam by Tomboy out of Tesane, by Whisker, with Colt by Man-at-Arms (foaled April 13), covered by Y. Trumpeter.

28. MAID SERVANT, b., 10 yrs. old (dam of Sentry and a Colt by Trumpeter), by Vedette out of Glenochty, by Plenipotentiary out of Glenlui, covered by Man-at-Arms.

28*. Bay FILLY by Man-at-Arms out of Maid Servant (own sister to Sentry). Foaled March 20.

29. MARCHIONESS MARIA, br., 6 yrs. old, by Colsterdale out of Princess Maud, by Touchstone, her dam Princess Alice, by Liverpool out of Queen of Trumps, by Velocipede, with Colt by Camerino (foaled February 21), and covered by him again.

30. MERRY MAY, b., 17 yrs. old, by Orlando out of Martha Lynn (Voltigeur's dam), by Mulatto out of Leda, by Filho da Puta, and covered by Y. Trumpeter.

30*. COLT by Camerino out of Merry May. Foaled January 24.

31. MISS GOLDSCHMIDT, br., 24 yrs. old, by Ion, her dam Jenny Lind, by Touchstone out of Malibran, by Whisker, covered by Man-at-Arms.

32. MISS WINKLE, b., 7 yrs. old, by Newminster out of The Belle, by Slane, her dam Miss Fairfield by Hampton (Grey Tommy's dam), by Comus, with Colt by Camerino (foaled March 8), covered by Y. Trumpeter.

33. MODENA, ch., 9 yrs. old, by Rataplan out of Ferrara,

by Orlando out of Iodine, by Ion, covered by Man-at-Arms.

34. MONTANA, b., 6 yrs. old, by Rataplan out of Mountain Flower (foaled in 1849), by Ithuriel out of Heather Bell, by Bay Middleton, with Filly by Promised Land (foaled February 26), covered by Y. Trumpeter.

35. MRS. WOLFE, b., 7 yrs. old, by Newminster out of Lady Tatton, by Sir Tatton Sykes, her dam Fair Rosamond, by Inheritor, and covered by Camerino.

35*. B. FILLY by Camerino out of Mrs. Wolfe. Foaled January 16.

36. MYRUS, b., 5 yrs. old, by Stockwell out of Lelia, by Melbourne out of Meeanee, by Touchstone out of Ghuznee, by Pantaloon, with Filly by Man-at-Arms (foaled March 11), covered by Y. Trumpeter.

37. NUDITY, ch., 6 yrs. old, by Crater out of Petticoat, by Pantaloon out of Camp Follower, by the Colonel out of Galatea, by Amadis, with Colt by Man-at-Arms (foaled March 1), covered by Camerino.

38. PRUDENTIA, br., 5 yrs. old, by Blair Athol out of Prudence (dam of Socrates, Benefactor and Astolfo), by Voltigeur, her dam Gossamer, by Birdcatcher out of Cast Steel, by Whisker, covered by Y. Trumpeter.

39. RAPIDAN, br., 12 yrs. old, by Beadsman (winner of the Derby), her dam Miami (winner of the Oaks), by Venison out of Diversion, by Defence, covered by Man-at-Arms.

39*. B. COLT by Man-at-Arms out of Rapidan. Foaled January 18.

40. RIBBON, b., 12 yrs. old (dam of Harmless), by Rataplan, her dam Lady Alicia, by Melbourne out of Testy, by Venison, covered by Camerino.

41. RIFLE PIT, ch., 13 yrs. old, by Rifleman out of Patch, by Sleight-of-hand, her dam sis. to Grey Momus, by Comus, covered by Camerino.

42. ROSELEAF, br., 9 yrs. old, by Gunboat out of Creeping

APPENDIX I.

Rose, by Surplice, her dam Rose of Cashmere, by Bay Middleton out of Moss Rose (sis. to Velocipede), with Filly by Camerino (foaled April 19), covered by Man-at-Arms.

43. SECRECY, b., 9 yrs. old, by Trumpeter out of Fairy, by Hermit out of La Femme Sage, by Gainsborough, grandam by Whisker, with Colt by Camerino (foaled May 26), and covered by him again.

44. SENSELESS, b., 19 yrs. old (dam of Pelios, Vacuum, Deceiver, and Senseless colt), by Bay Middleton out of Vanity, by Slane out of Breastgirth, by The Saddler, with Filly by Promised Land (foaled March 10), covered by Y. Trumpeter.

45. CRACK SHOT, ch., 7 yrs. old (sis. to Marksman), by Dundee out of Shot, by Birdcatcher out of Wasp, by Muley Moloch, covered by Camerino.

46. SILVIA, br., 11 yrs. old, by Vedette out of Sylphine, by Touchstone out of Mountain Sylph, by Belshazzar, covered by Y. Trumpeter.

47. SPOILT CHILD, 7 yrs. old (Naughty Boy's dam), by Wild Dayrell out of Bessie Bee, by Sweetmeat, her dam Lady of Littlecote, covered by Man-at-Arms.

48. SPELLWEAVER, b., 6 yrs. old, by Newminster out of Shamrock, by Y. Priam out of Miss Bucktrout, by Perion or Tomboy, grandam by Figaro, with Colt by Camerino (foaled March 11), covered by him again.

49. TATTOO, b., 15 yrs. old (dam of Vesuvius), by Rataplan out of Fandango, by Touchstone out of Sequidilla, by Sheet Anchor, with Filly by Camerino (foaled April 11), covered by Y. Trumpeter.

50. THE BANK, b., 7 yrs. old, by Voltigeur out of Dividend (dam of Rouble), by Auckland out of Diploma, by Plenipotentiary, with Filly by Camerino (foaled May 11), covered by Y. Trumpeter.

51. TISIPHONE, b., 14 yrs. old (dam of Alice), by Orlando out of Torment, by Alarm, grandam by Glencoe out of Alea, by Whalebone, covered by Y. Trumpeter.

28—2

51*. B. FILLY by Camerino out of Tisiphone. Foaled March 15.

52. VALLATION, br., 12 yrs. old (dam of Valuer), by Vedette, her dam Palmistry (dam of St. Giles), by Sleight-of-hand out of Mystery's dam, by Lottery, covered by Camerino.

52*. COLT by Man-at-Arms out of Vallation. Foaled January 25.

53. VERITY, br., 10 yrs. old, by Vedette out of Vera, by Touchstone out of Muscovite's dam, by Camel, covered by Y. Trumpeter.

53*. Ch. FILLY by Man-at-Arms out of Verity. Foaled January 7.

54. VESTMENT, ch., 7 yrs. old, by St. Albans out of Nettle, by Sweetmeat, her dam Wasp, by Muley Muloch, grandam by Emilius, covered by Y. Trumpeter.

55. VILLAGE MAID, b., 15 yrs. old (dam of Ploughboy and other winners), by Stockwell, her dam Minx (sis. to Melbourne), by Humphry Clinker, grandam by Cervantes, covered by Y. Trumpeter.

55*. B. COLT by Man-at-Arms out of Village Maid. Foaled January 26.

(3) YEARLINGS.

56. COLT, b., by Man-at-Arms out of Myrus. Foaled April 9.

57. COLT, br., by Man-at-Arms out of Prudentia. Foaled May 6.

58. COLT, br., by Promised Land out of Ribbon. Foaled March 25

59. COLT, b., by Camerino out of The Bank. Foaled March 29.

60. COLT, br., by Delight out of Delilah. Foaled February 7.

61. COLT, br., by Man-at-Arms out of Crack Shot. Foaled March 5.

APPENDIX I.

62. COLT, br., by Delight out of Castanette. Foaled April 10.

63. COLT, b., by Promised Land out of Nudity. Foaled March 9.

64. COLT, br., by Camerino, dam by Surplice out of Bellona. Foaled in April.

65. COLT, b., by Man-at-Arms out of Hopeless. Foaled May 13.

66. FILLY, ch., by Trumpeter out of Tattoo. Foaled April 24.

67. FILLY, b., by Camerino out of Berenice. Foaled March 11.

68. FILLY, ch., by Camerino out of Merry May. Foaled January 24.

69. FILLY, b., by Camerino out of Mrs. Wolfe. Foaled February 7.

70. FILLY, b., by Man-at-Arms out of First Lady. Foaled April 1.

71. FILLY, b., by Camerino out of Pugilistic Art. Foaled February 15.

72. FILLY, b., by Camerino out of Australia. Foaled February 22.

73. FILLY, br., by Man-at-Arms out of Spoilt Child. Foaled March 27.

74. FILLY, b., by Camerino out of Crossfire. Foaled January 16.

75. FILLY, br., by Man-at-Arms out of Silvia. Foaled May 4.

76. FILLY, bl., by Man-at-Arms out of Montana. Foaled March 15.

77. FILLY, bl., by Camerino out of Burgas. Foaled January 25.

78. FILLY, ch., by Man-at-Arms out of Miss Bell. Foaled March 12.

79. COLT, b., by Camerino out of Araby's Daughter. Foaled in March.

80. COLT, br., by Toxophilite out of Village Maid. Foaled February 23.

81. COLT, br., by Camerino out of Vallation. Foaled January 31.

82. COLT, b., by Camerino out of Miss Winkle. Foaled April 2.

83. COLT, b., by Man-at-Arms out of Cracovienne. Foaled March 20.

84. COLT, b., by Man-at-Arms out of Little Jemima. Foaled May 6.

85. COLT, b., by Camerino out of Rifle Pit. Foaled March 6.

86. COLT, br., by Man-at-Arms out of Chic. Foaled March 19.

87. COLT, b., by Camerino out of Verity. Foaled February 9.

88. COLT, ch., by Man-at-Arms out of Cochineal. Foaled February 5.

APPENDIX II.

LIST OF STALLIONS.

THOROUGHBRED SIRES OF 1883, AND STALLIONS SERVING IN 1884, IN ENGLAND.

Abbot (The)
Albert Edward
Albert Victor
All Right
Altyre
Apollo
Argyle
Arrow (The)
Ascetic
Astrologer
Attalus
Avontes
Awfully Jolly (a Barb)

Balfe
Balquhidar
Banbury Bun
Banneret
Barb (A)
Barbillon

Barcaldine
Barefoot
Beau Brummel
Beauchamp
Beauclerc
Beaudesert
Bedworth
Bellerophon
Ben Battle
Bend Or
Berserker
Bertram
Billy
Blandford
Blantyre
Blue Ruin
Bold Dayrell
Bold Marshall
Bookworm
Break of Day
Bridgeford

Brigand
Brilliant
Bruar
Bruce
Bugler
Bustard

Caerau
Callistos
Camballo
Cardinal York
Carnelion
Castalian
Castlereagh
Cathedral
Cedric
Cerberus
Change
Charibert
Chevron
Chilblain

Childeric
Chippendale
Chypre
Clanronald
Claremont
Cœruleus
Colombo
Coltness
Conductor
Consternation
Controversy
Corporal
Creancier (bred in France)
Cucumber
Cultivator
Cymbal
Cyprus

Dan Godfrey
Dart
Death or Glory
D'Estournel
Discord
Disturbance
Doncaster
Don Fulano (bred in America)
Dr. Temple
Drumhead
Duke (The)
Dutch Skater

Earl (The)
Earl Clifton

Earl of Dartrey
Edward the Confessor
Escort
Ethelred
Ethus
Exminster

Farnese
Fernandez
Fetterlock
Feuris
Fiddler
Fitz James
Flying Birdcatcher
Forerunner
Foxhall (bred in America)
Friar Tuck

Galopin
Genuine
George Frederick
Gladstone
Glendale
Golden Horn
Goodwood
Greenback
Grey Palmer
Gunboat

Hackthorpe
Hagioscope
Hampton
Happy Land

Hermit
Heron
Highborn
Hilarious
Hillingdon
Holmby
Horse (half-bred)
  "   "
  "   "
  "   "
Hydromel

Incholm
Iodine I.
Isonomy

Janie Croft
Jarnac
Jock of Oran
John Davis
John Day
Jolly Friar
Julius Cæsar
Jupiter

Kars (an Arab)
Khamseen
King Alfred
Kingcraft
King Harold
King Lud
King of Trumps
Kisber (bred in Austria)
Knight of the Garter

APPENDIX II. 441

Knight of the Launde	Melville	Pepin le Bref
Kouch (an Arab)	Mereworth	Peppermint
	Mexico	Peregrine
	Miltiades	Pero Gomez
Lancastrian	Mirmillo	Peter
Landmark	Misenus	Petrarch
Landseer (an American horse)	Miser	Petronel
	Montez	Picnic
Lecturer	Moriturus	Playfair
Liddesdale	Mornington	Plebeian
Lionel	Mosquetaire	Posta (an Arab)
Little Wonder	Mr. Winkle	Poulet (bred in France)
Londesborough	Muncaster	
Lord Clyde	Munden	Prince
Lord Derby	Mustapha	Prince Charlie
Lord Glenlyon		Prince George
Lord Gough	Napsbury	Privateer
Lord Lyon	Navigator	Privilege
Lord Malden	Neophyte	Professor
Lowlander	Newfield	Pursebearer
Lucebit	New Holland	
Lumley	Newry	Queen's Messenger
Lurgan	Norwich	Queen's Herald
	Not Out	Quits
Macaroni	Nuneham	
Macaroon		Rattle
Macgregor	Omega	Redcap Sly
Make Haste	Onslow	Restless
Mappleton	Osric	Retreat
Marden	Our John	Reveller
Marmion	Outfit	Reverberation
Mask		Richelieu
Master Kildare	Pedometer	Robert the Devil
Master Waller	Pellegrino	Rococo
Melton	Pell Mell	Rosbach

Rosicrucian	Strathavon	Van Amburgh
Rosebery	Strathearn	Vanderdecken
Rostrevor	Strathmore	Vanguard
Rotherhill	Struan	Victor Chief
Royal George	Stylites	Victorious
Ruperra	Sugar Plum	
	Sweetstock	Wallenstein (bred
Scamp	Sykes	in America)
Scottish Chief	Syrian	Waverney
See-Saw		Wellingtonia
Sefton	Teviotdale	Wenlock
Shifnal	Thorwaldsen	Westbourne
Siderolite	Thunder	Westcombe
Silurian	Thunderbolt	Wild Oats
Silvester	Thuringian Prince	Wild Tommy
Sir Bevys	Thurio	Winslow
Sir Frederick	Tiber	Wisdom
Skylark	Tibthorpe	
Snail (The)	Tomahawk	Xanthus
Speculum	Touchet	Xenophon
Springfield	Town Moor	
St. Anthony	Trappist	
St. Peter	Troll	Young Trumpeter
Standard	Tunis	
Star of the West H B	Tyndale	Zanzibar
		Zeal
Statesman	Uncas	Zealot
Sterling		Zeltinger
Storm Signal	Valour	Zephyrus

INDEX.

ABNORMAL presentations, 245
Abortion, causes of, 114
Absence of authoritative modern teaching, 6
Accommodation for mares and yearlings, 379
Acre of land, time taken in ploughing an, 283; three-quarters of only ploughed in a day, 297; farmers' ignorance of distance travelled in ploughing an, 308; exact calculation of the same, 310
Acreage under crop in Great Britain, cost of, 322
Action affected by exercise, 103; good action inherited, ib.; should be looked to in selecting brood-mares, 181
After-birth, the, 245
After covering, treatment of mare, 241
Age, the, of mares and stallions, 166; curious facts relating to, 219; respective ages in mating, 220; age to put a stallion to the stud, ib.
Agility and Mandragora, purchase of by Mr. Gee, 178
Agitators, agricultural, 331
Agricultural horses, how to breed, 277; thoroughbreds for farm work, 286; half-breds for farm work, 275-311
Agricultural work and wages, 292, etc.; in the eastern and southern counties, 299; estimates, 300; lack of progress and improvement in, 319; statistics of, in Great Britain, 322
Agricultural labourer, the, 327, etc.
Agriculturists, masters and men, better times for, 303
Aim in writing, my, 2

Alice Hawthorn, 184; her racing and stock, 212
Allie Slade, an instance of want of courage, 128
Alvediston, my stud at, 5; list of, 430; lime-trees at, a cause of mares slipping, 116; clock-tower at, its cost, 388
Americans, the, and farm-horses, 308
American trotter, the, as compared with the English race-horse, 167; account of, 343, etc.
Ancient horse-racing, 42
Annual profit to be derived from twenty-four half-bred mares, 280
Arab, the, as he is, 76; covering fees of noted Arabs, ib.
Arabia not the original home of the horse, 33
Arabian horses, first introduction of into England, 45, 52; the Duke of Newcastle on, 45; noted sires, 48; not successful on the course or at the stud, 67; incapable of improvement, 89
Arab strain, real value of, 63-79; failure of since *Eclipse's* time, 72; protest against re-introduction of, 78
Aspect, the, of the stud farm, 365; Cecil on, 379
Ass, the, 355-363
Athelstan's, King, 'running horses,' 43; cross got from, 53
Attila, trial of, at Newmarket, 346
Auroch, the, and interbreeding, 403
Authors' mistakes, some, 392

Babraham, value and description of, 77
Bad smells, effect of, on pregnant mares, 115

Bad temper, 120-127; a fault overlooked by breeders, 417
Balaklava, the charge at, 265
Bald Charlotte, her performances, 68
Barbs and Arabians, first introduction of into England, 45, 52; descendants of, 59
Bay Middleton, his racing, 409
Bear-baiting, 41
Beeswing, 184; her racing and stock, 211
Bell on the Arabian horse, 33
Belladrum, his roaring cured in South Africa, 133
Bend Or, his pedigree, 162
Bentinck, Lord George, his breeding experiments, 164; his purchase of *Octaviana* and *Crucifix*, 178
Bison, the, and interbreeding, 403
Bitches, lining of, 234
Black cart-horse, the old, 340
Black-thorn hedges a mistake, 370
Blenkiron, Mr., and breeding from one's own stallion, 216
Blood, the, its character no test of breeding, 406
Boon to agriculturists, a, 278
Bowels, regulation of the, 248; attention to stallion's diet in regard to his, 260
Boxes, the, for mares and yearlings, construction of, 379; Cecil on aspect of, 380; roof, walls, interior, 332, etc.; for stallions, 388; various kinds, *ib.*; yards around, 390
Brains of animals, 'Stonehenge' on, 403; brain-disease in a horse, 405
Breeder, every man his own, 25
Breeders, how to check the evils caused by ignorant, 426
Breeding as a national question, 12-25; much written about and discussed, 18; for the army service, 19; opinions on, 20; our national advantages for, 24; first royal encouragement given to improvement in, 45; eligible sites for, 102; purity of blood essential in for producing good stock, 137; tendency to revert to parent types in, 138; Darwin on, 139; Youatt on, 140; evils of in-and-in breeding, 141; crossing. 145; general principles of selection for, 149-169; breeding from unsound horses, 152; suitability of sire and dam in, 158; diversity of opinion on, 160; in-and-out breeding, *ib.*; why few mares produce good horses, *ib.*; experiments in of Lord George Bentinck, 164; in selection of mare, 171; distinguished or young mares the best for, 172; relationship to good horses a great point in, *ib.*; luck in, 176; in selection of stallion, 189; curious facts in, 195; influence of sire and dam upon respectively, 200; suitability of sire and dam in, 202; age for, *ib.*; size best for, 203; judicious crossing in, 205; remedying too great or little size in, 207; mares breeding best from particular stallions, 208; racing and, 210; objections to breed from one's own stallion, 215; examples for and against the practice, 216-218; curious facts relating to age in, 219; of hunters, 267-270; of hunters for heavy weights, 272; of cart-horses, 275; of carriage-horses, 277; of troop-horses, *ib.*; estimates of, for farmers, 279; of asses, 359; improvements in breeding asses, 361; the stud farm, 364-391; obsolete works on breeding to be avoided, 403; blood not a test of, 406; numbers of thoroughbreds, 413; reform imperative in, 428.
British labourer, the, 319, etc.
British wild-cattle at Chillingham, the, 98
Broeck, Mr. Ten, and late breaking in, 409
Brood-mares and foals (see MARES, FOALS, etc.); at Alvediston, 430
Buccaneer sold abroad, 423
Buffon on interbreeding of goats and sheep, 394; his error exposed, 395, etc.
Bugler, price refused for, 121
Buildings, the stable, materials for, 381; the roof and walls, 382; the interior, *ib.*; various materials for, 388
Bull-dog and greyhound, cross between, 143
Bullocks in the stud paddocks, 369
Byerly Turk, the, as a sire, 55

Calculation of waste of time, 301
Canezou, experiment with, 206
Cape of Good Hope, roaring unknown at, 132

INDEX.

Care needed during pregnancy, 114
Carters and ploughmen, their work, 291; less work and more wage for, 292; idleness of, 296, 302
Cart-horse, the, rate of walking, 282; question whether better or worse than in former times, 288; substitutes for, 291; working hours of the, 291; on the road, 304; no improvement in, 307; inferiority of, 314; comparison of with other draught-horses, 324
Castrating-knife, argument for wider use of, 221-224
Castration advocated to improve our breed, 419
Cat, story of, and the *Godolphin Arabian*, 58
Cataract, decrease of, 130
Cattle, weight of, 101; and sheep, English breeds the best, 97
Causes of abortion, 114; of deterioration of size in animals, 353
Cecil, his 'Stud Farm,' why unsatisfactory, 7; on the effects of an inferior cross, 195; on aspect of stables and paddocks, 379
Central Asia, the cradle of the horse, 35
Cervantes mare, the, purchase of, 177
Chanticleer, descent of his colour, 146
Chapman type, the, 342
Chillingham cattle, the, 98
Choice of the mare, 185
Claverhouse in the plough, 286
Cleveland Bays, the, 341
Climate, effect of, 89; and other influences, 95-105
Climatic effects on horses, 96; on dogs, *ib*.
Coaches, rate of speed of, 313
Coach-horses, rate of speed of, 313; loads drawn, 314; comparison of with other draught-horses, 324
Cob, the, 350, etc.
Cock-fighting, decline of, 40
Coercion as to gelding advocated, 417
Coffin mare, the, 46
Cole, Mr., anecdote of, relative to farmwork, 307
Colour transmitted through generations, examples of, 145
Comparison of draught-horses, 312-325; table of, 324
Comus at the stud, 267
Conception, moment of, 229

Condiments a mistake, 249
Conformation an essential point in selecting sire and dam, 153
Consanguineous intercourse and its results, 136-145
Consequence, the, of the horse, 326-335
Constable, Sir Richard, and the *Cervantes* mare, 176
Constipation in the foal, how to relieve, 246; in the stallion, 260
Constitution, consideration of in selection, 184
Corn, storage of, 386
Cost of breeding horses on a farm, 277; estimates of, 279
Costermongers' donkeys, 358
Covering the mare, best time for, 234; fees, 420; fees of noted Arab sires, 77
Cows' milk for foals, 249   [145
Crab and his offspring, the grey colour,
Craven, Mr. W. G., his letter to the *Daily Telegraph*, 19
Cromwell, and racing, 46
Cross, the thoroughbred, in hunter and troop-horse, 263-274
Cross-bred stallions and mares, 161
Crosses, original, 52; between mare and zebra, 197; between mare and quagga, *ib*.; between horse, zebra, quagga, onager, and ass, 356
Crossing of breeds, 145; mare spoiled for life by one inferior cross, 195; effects on future offspring, *ib*.
Croydon Board of Guardians, the, and expenditure, 333
Crucifix, her temper, 122; her racing and stock, 212
Cultivation in Great Britain, 323
Curwen's Bay Barb, arrival of, 56

Dale, Thomas, feat of donkey-riding, 358
Darley Arabian, the, as a sire, 55
Dartmoor ponies, 353
Darwin, Charles, on breeding, 139; on crossing and fertility, 140; on selection for mating, 153
Davis, Mr., a veteran huntsman, 16
Dead foals, bearing, a cause of sterility, 111
Dearth of good horses in the West, 23
Decline of cruelty in sport, 41
Defects of stallions now at the stud, 421

Defective sight, 128; blindness hereditary, 129; Mr. Sadler's loss thereby, *ib.*; decrease of cataract, 130
Defence, his blindness, 129
Defenceless, stock out of, 209
Delilah, her worthless stock, 175
Delivery of the mare, 245
Depreciation in value of cattle and horses, 278
Derby, Earl of, his experiment with *Canezou*, 206
Derby winners got by sires not belonging to their dams' owners, 218
Desirable qualities to obtain in half-breds and draught-horses, 338
Devonshire, revival of breeding in, 23; thoroughbred stock in, 266
Devonshire pack-horse, the, 341
Dexter, extraordinary performance of, 347
Diarrhœa in horses, 261
Discussion of possibility of interbreeding between the goat and sheep, 394-401
Disease, the stallion's power to transmit, 188
Diversity of opinions on breeding, 160
Domestic ass, the, possible improvement of, 358; instance of speed and endurance of, *ib.*; selection of for breeding, 360
Doncaster, his pedigree, 162
Donkey, the English, the most inferior of his race, 360 [385
Doors of the stables, construction of,
Drains in the stables, 384
Draught-horses, comparison of, 312-325; comparative table of speed and work of, 324; and trotters, 336-349; sundry kinds of, enumerated and described, 339, etc.
Dray-horse, the breeding of, 339
'Druid' on the Chillingham cattle, 98; on sheep, 100; anecdote related by, 346
Dung-pits, effluvia from, a cause of abortion, 115

Early maturity and early decay, 408
Eastern counties, farm-work in, 299
Eastern sires, first importation of, 48
Eclipse, 60; his performance, 61; compared with *Touchstone*, 87; at the stud, 166; descendants of, 215; lameness of, 256; late appearance of on the turf, 408

Edwards, Mr., William IV.'s trainer, 49
Effect of climate, 89; of consanguineous intercourse, 141
Egypt the early home of the domesticated horse, 31
Elcho's dam, stock out of, 209
Eligible sites for breeding, 102
Elmsthorpe and brain-disease, 405
Enclosing the stud paddocks, 370
English donkey, the most degenerate of his kind, 360
English racing, beginning of, 41; horses appreciated abroad, 91; racehorse and American trotter compared, 167
Entre Nous, her temper, 122
Equidæ, the, 28
Errors and fallacies, some, 392-411
Estimates of cost and profit of breeding half-breds on a farm, 279; of farmers' profits, 281; of ploughing, 283
Examples of good early sires, 70; of early failures, 71
Exclusion from 'The Stud-Book,' 429
Exercise in relation to action, 103; as developing good action, 182; of yearlings, 251; of the stallion, 255; want of, a cause of mortality among stallions, 258; yard for the stallion, 390
Existing works on breeding not satisfactory, 7, 12
Experience, my own, 4
Export of English horses, 91
Extremes of size in breeding, 203

Failure of Oriental cross since *Eclipse's* time, 72; of large stallions, 192
Failures, good-looking, 174
Faint-heart, hereditary and ruinous, 127
Fallacies and errors, 392-411
Falmouth, his temper, 124
Farm, half-bred on the, 275-289
Farmers, possible profits of, 281; gains from using half-breds in the plough, 284; gains from employing a stableman, 287; paying for work never done, 297; ignorance of relative amounts of work, 306; want of energy among, 321
Farquharson, Mr., a veteran of the hunting-field, 15; and late breaking in, 409

INDEX. 447

February, the middle of, best time for covering, 231
Feeding, requisite, for mares with foal, 244; for foals, 247; with corn, *ib.*; of plough-horses, 299
Feet, the foal's, 243; the yearling's, 251; the stallion's, 254; evil effect of shoes, *ib.*; attention to, 255
Fertility, Darwin on, 141
Field, the, article in on thoroughbred sires, 265
Fifteen first-rate thoroughbred stallions enough to keep annually, 425
Final suggestions, some, 412-429
First-born foals, good, 202
First trace of a cross in the British horse, 39
Fitz-Hardinge, Lord, and breeding from young horses, 280
Flora Temple, extraordinary performance of, 347
Flying Childers, birth of, 55; speed of, 56
Flying Duchess, stock out of, 209
Foaling, time of, 244
Foals, scouring of, a sign of mare being at use, 235; treatment of, 242-252; treatment of hunter class of, 271; overfeeding of, 349
Foreign sales of our best horses an evil, 423; winners of the Derby, 50
Fox, the, past and present, 264
Fox-hunting and hunters, 15
Frail, Mr., and the yearling race at Shrewsbury, 42
French coercive measures, 417
Fright a cause of abortion, 115
Furniture-vans, distance travelled by, weight, speed, 316
Furze-hovel, the, 389

Galloway and pony, the, 350-355
Gaudy and her offspring, transmission of colour, 145
Gelding-knife, the, more extensive use of needed, 414; Government should enforce use of, 417
Geldings and the Queen's Plates, 221; useful on the turf, 416
George IV. and his horses, 288
Gestation, period of, 232
Gilbey, Mr. Walter, his opinion on breeding, 20; on the shire horse, 339
Glee, good example of a small broodmare, 180

Goat and ewe, Buffon's statement as to interbreeding of, and my refutation thereof, 394, etc.
Godolphin Arabian, the, 55; his merits and produce, 57; story of black cat and, 58; his value, 66
Goldsmith's statement as to interbreeding of bison and cattle, disproof of, 403
Good-looking mares failures at the stud, reason and examples, 174
Good stallions only should be kept, 423
Goodwood Cup, the, of 1830, 49
Grace Darling, purchase of, 177
Grandfather, my, and farm-work, 306
Grass on the stud farm, 365; feeding off of, 369
Grey Friar, perpetuation of his colour in descendants, 147
Greyhound, the, examples of effects of exercise on, 104; cross of with bulldog, 143

Habitat, original, of the horse, 31
Hack, the, 350, etc.
Half-and-Half's stock, 267
Half-bred, the, on the farm, 275-289; *versus* the shire-horse, 290-311
Hampshire horses, 368
Hampton Court stud, the, founding of, 48; paddocks at too small, 370
Handling the foal, 242
Hare-hunting, 17
Harfit, Mr., and speed of van-horses, 316
Hartwig, Dr., on labour in Mexico, 320
Hawley, Sir Joseph, and breeding from one's own stallion, 217
Hay, Major W. E., on the kiang or onager, 357
Hay, Mr. Delisle, on piece-work in New Zealand, 294
Hay-racks, place for in the stable, 384
Hay, storage of, 387
Hayter, Messrs., on interbreeding sheep and goat, 399
Health of the stallion, the, in respect of selection, 188
Heavy-weight carriers, how to breed, 272
Hedges enclosing paddocks, 370; planting of, 373 [387
Helpers' sleeping-room in the stable,
Henry VIII., his horses, 44

THE HORSE.

Herod, his running and stock, 59
Highflyer, late appearance of on the turf, 408
History of the horse, 26-39
Hobbling mares, expediency of, when covered, 228
Horse, the, history of, 26-39; antiquity of, 27; extinct species of, *ib.*; zoological classification of, 28; general description of, 29; Biblical account of, 31; original habitat of, *ib.*; Egyptian use of, 32; origin of not proved to be Arabian, 33; King Solomon's horses, *ib.*; Mohammed and, 35; cradle of in Central Asia, *ib.*; mention of by Josephus, 36; in Britain, 36; used during Cæsar's invasion, 37; British war-chariots, 38; first cross introduced, 39; at Smithfield fair, 43; Henry VIII. and, 44; improvement in by Charles II., 46; Eastern sires, 48; earliest crosses, 52; Roman cross, 53; Athelstan's cross, *ib.*; Spanish strain, 54; Arabian strain, 55-62; real value of the Arab strain, 63-79; improvement of the modern horse, 80-94; spread of the English breed abroad, 91; in New Zealand, 94; temper of the horse, 119-127; crossing of, 136-148; general principles of selection, 149-169; treatment of the stallion, 253-262; hunter and troop-horse, 263-274; speed compared with hounds, 268; kind of best for farmers, 275; carriage-horse, how to breed, 277; troop-horses, *ib.*; value and cost of, *ib.*; rates of walking, 282; thoroughbreds for farm-work, 286; working hours of cart-horses, 291; on the road, 304; comparison of draught-horses, 312-325; consequence of the horse, 326-335; the basis for agricultural reform, 334; draught-horse and trotter, 336-349; various kinds of horse, 339; dray-horse, *ib.*; Suffolk Punch, 340; Cleveland Bay, 341; Devonshire Pack-horse, *ib.*; Chapman type, 342; small kinds of, 350-355; crosses with zebra, ass, etc., 356; where large and small thoroughbreds are reared, 368; obsolete works on to be avoided, 403; late breaking in of, 407; longevity of, 410

Hounds faster than horses, 268; and hunters and foxes, past and present, 264
Houses on stud-farm, 378
Housing and feeding of brood-mares, 241
Hovels or boxes, construction of, 379; Cecil on aspect of, 380; the stallions', 388; various building materials, *ib.*; the furze-hovel, 389; yards around the, 390
How to breed hunters, 270
Hunter and troop-horse, the, 263-274
Hunters, plenty of good sires for, 265; how to select for breeding, 269; several good thoroughbred sires, 337
Hunting in Persia, 13; among the Romans in Britain, *ib.*; under Canute, 14; under the Normans, *ib.*; recommended by James I., *ib.*; in relation to health, 15; to longevity, *ib.*; of the stag, 16; of the hare, 17; the sport of princes, 18; fox and stag-hunting have led to development of speedier and more enduring horses, 343
Huntsmen, octogenarian, 16
Hybrid crosses, 197
Hypothesis, the, of blood or blood-vessels being a test of breeding discussed, 407

Illustrations of paddock-railings, 374, 375
Improved quality required in our horses, 24
Improvement of horses, 3; of the modern horse, 80-94; of coach and post horses, 288
In-breeding a mistake, 144
Indications of coming delivery of mare, 244
Ineligible sites for stud farms, 367
Inferior stallions and winners, 427
Influences on breeding, 95-105
Injudicious selection, 21
Interbreeding of goat and ewe, discussion on, 394, etc.
Introduction, 1-11

Jack-ass, landlords should keep a good, 362
Johnson, Dr., on little increase in learning, 317; on the value of time, 327; on the acquisition of knowledge, 397

INDEX.

Judgment of horses not possible without trial, 156

Kiang or onager, Major Hay on, 357, 358

Labour and wages, difference of agricultural, 296 ; in Mexico, 320
Labourers, why the best leave the farm, 328
Lameness, 151 ; of stallions, 256
Large and small thoroughbreds, where respectively reared, 368
Largesse, a good-looking failure, 174
Late foaling, 236
Like vascular structure in all horses, 407
Lime-trees at Alvediston, cause of mares slipping, 116
Little Tom, his running, 121
Localities of thoroughbred stud farms, 366
Loft, the stable, 386
Longevity in relation to early breaking and high feeding, 410
Lonsdale, Lord, ploughing with thoroughbred mares, 286
Luck in breeding, instances of, 176

Maid of Palmyra, her racing and stock, 211
Mamhead paddocks, the, 368
Management, present systems of good ones, 410
Mandragora and *Agility*, purchase of by Mr. Geo, 178
Mangers, construction of, 383
Mare, the, dangers of bad temper in, 124 ; age to put to the horse, 166 ; selection of, 170-186 ; breeding of, 170-179 ; shape and size of, 180 ; action of, 181 ; constitution of, 184 ; speed and stamina of, 185 ; effect on of inferior stallion, 195 ; the thoroughbred, 199-212 ; influence of on the foal, 201 ; predilection of for some particular stallion, 208 ; bad racers good breeders, 210 ; pedigree the test, *ib.* ; instances in point, 211 ; time for covering, 228-231 ; hobbling of when being served, 228; repetitions of service of, 230, 234 ; introduction of to the horse, 231 ; period of gestation in, 232 ; stinting of, 233 ; turning of, 235 ; late foaling of, 236 ; a year's rest for, *ib.* ;
treatment of, 238-252 ; when unfit for breeding, 240 ; hunting mares not put to the stud till late in life, 270 ; half-bred for farm work, 276 : twenty-four mares, an estimate, 279, etc. ; in the paddock, 369 ; numbers of thoroughbreds at the stud, 413 ; too many bred from, 415 ; every thoroughbred mare covered by Eastern stallion or half-bred should be excluded from 'The Stud-Book' ; list of brood-mares at Alvediston, 430

Margellina, Mr. Wreford's successful mare, 173
Marie Stuart, sale of, 178
Markham Arabian, the, purchase and failure of, 45, 55
Marske, the sire of *Eclipse*, 61
Master's house, proper position of on stud farm, 378
Matchem, his running, 59
Mating, 227-237 ; for breeding hunters, 269
Melbourne and in-breeding, 144
Messenger, the original sire of American trotters, 344 ; at the stud, 347
Miss Twickenham, her racing and stock, 211
Mortality among stallions, 257 ; causes of, *ib.* ; remedy for, 259
Mouche, Mr. Wreford's unsuccessful mare, 174
Mule, the, 357
Musket, foreign sale of, 423

National cultivation, cost of, 322
National question, breeding as a, 12-25
Need of practical knowledge, 6
Newcastle, stallion show at, 22
New Forest pony, the, 353
'Newmarket and Arabia,' Upton's, 64
Newminster, his lameness, 256
Newton-Fellowes, Mr., his coaching team, 266
New Zealand, progress of racing in, 93 ; well-bred horses in, 94 ; Mr. Delisle Hay on work in, 294
'Nimrod's' account of savage horses, 418

Oats, storage of, 386
Octaviana and *Crucifix*, purchase of, 178
Octogenarian fox-hunters, 15
Old-fashioned hunters and hounds, 264

29

Old Forester, his temper, 418
Omega, at the stud, 267
Omnibus horse, speed of, and load drawn, 315, 324
Onager, the, 355 ; Major Hay on, 357
Origin of ponies, the, 353
Ormonde, whence he inherited roaring, 131 ; his pedigree, 162
Our breeds the best, authorities in support of, 73
Ouseley, Sir Gore, his cross between an Arab and a zebra, 197
Overfeeding of stallions, 258 ; of foals, 349

Paddocks, the, general character of, 368 ; feeding off, 369 ; size of, 370 ; enclosing of, *ib.*; plantations round, 371 ; railings for, the right and wrong sort, 374, 375 ; shape of, 375 ; trees within, 376
Paradigm, instance of a good broodmare, 172 ; her racing and stock, 211
Paradox, a modern savage, 419
Pastures, various, of stud farms, 366
Paucity of good horses bred annually, 422
Pauperism, how to extinguish among agricultural labourers, 331
Pavement of stables, the, 385
Payment of wages, system of, and the poor laws, 332
Pedigree a safeguard in selection, 157 ; a test of soundness, 184
Pembroke, Earl of, as a hunting man, 17
Petronel, pedigree of, 163
Physic, 250
Piece-work and time-work, 293, etc.
Pig, my Russian, 90 .
Pigeons, all kinds of derived from blue rocks, 137
Placenta, the, 245
Place's White Turk, importation of, 46, 55
Plan of this book, 8 ; my, to improve our breed, 425
Plantations round paddocks, 371
Ploughing, distance travelled and time spent in, 283 ; with thoroughbreds, 286 ; calculation of distance made and time taken in, 310
Ploughman and stable-keeper, the, 287
Poisonous trees, 373

Pony and galloway, 350-355 ; feat of riding of my father and Mr. Dilby, 351 ; *Sir Teddy's* marvellous performance, 352
Portman, Viscount, as a hunting veteran, 15
Portsmouth, Lord, and breeding from thoroughbred sires, 266
Post-horses, speed of and load drawn by, 315, 324
Powney, Mr., and *Grace Darling*, 177
Practical man the best authority on practical matters, a, 402
Preakness, not ridden till eight years old, 409
Pregnant mares not known to be in foal, 230
Prehistoric remains of the horse, 26
Presentations, abnormal, 245
Prince Charlie, too large for the stud, 192 ; my opinion of, 424
Pritchard, Professor, on accidental causes of disease, 134 ; on selection, 153 ; on crossing, 206
Profits for farmers, 281, etc.
Progress of racing and its results, 40-50 ; of science and useful arts, 318
Purging, how to prevent, 261
Purity of blood essential for production of good stock, 137

Quagga, the, and cross with mare, 197
Quality needed, 24
Queen Anne and racing, 47
Queen's Plates, the, and admission of geldings to run for, 221

Race-horse, the, improved by Charles II., 46 ; excellence of the present, 83 ; foreign, 92
Racing, progress and results of, 40-50 ; rise of in England, 41 ; practised by the Greeks, 42 ; earliest record of in England, 43 ; in Smithfield and Hyde Park, *ib.*; in Elizabeth's reign, 44 ; under the Stuarts, 45 ; under Cromwell, 46 ; under Charles II., *ib.*; under Anne, 47 ; under the Georges, 48 ; at Goodwood in 1830, 49 ; of *Eclipse*, 60 ; progress of, abroad, 92 ; in New Zealand, 93
Radcliffe, Mr., and piece-work, 293

INDEX.

Radclyffe, Mr., of Hyde, a veteran of the hunting-field, 15
Radnor, Earl of, a veteran of the hunting-field, 15
Railings of paddocks, description and illustration of, 374, 375
Rain-water, storage of, 381
Rams and ewes, age for breeding, 203
Rapidan, indifferent stock of, 175
Reduction necessary in number of stallions, 413
Reforms imperative, 428
Refusing to cover, stallions, how to treat, 261
Respective ages of sire and dam, 202
Ribblesdale, Lord, his opinion of my Alvediston stud, 5
Roaring, hereditary tendency of, 130; lying dormant, 131; in *Ormonde*, *ib*.; prevention of, 132; effects of climate on, 133; *Belladrum* at the Cape, *ib*.
Robinson, Mr., lucky purchase of the *Cervantes* mare, 177; and physic, 250; and his short-horn bull, 259; and ploughing with thoroughbreds, 286
Roe, the, only instance of a thoroughbred mare once crossed by half-bred stallion afterwards getting a winner, 196
Roman cross, the, 53
Roof of stable-buildings, the, 382, 386
Rotation of animals to feed off the paddocks, 369
Rous, Admiral, why his book is unsatisfactory, 7; on gameness, 82; on present superiority of English race-horse, 83; wrong advice of in respect to paddock-hedges, 370
Royal mares, the, 47
Royal patronage of racing, 40-50
Russell, Mr. John, the veteran sporting parson, 16
Russian pig, my, 90

Sabin, Mr., and piece-work, 294
'Saddle and Sirloin,' 'Druid's,' 100
Sadler, Mr., loss of his blind mare, 129; in-breeding of game-cocks, and result, 142; curious tale of fowls' eggs and a raven, 143; his pregnant mare put to *Venison*, 230
Sancho Panza and his ass, moral of the tale, 363

Schism, and her inferior stock, 175
Scott, John, anecdote of, 346
Seasons for tree-planting, 371
Selection, general principles of, 149-169; Pritchard and Darwin on, 153; Sir Tatton Sykes' method of, 154; shape in, 153; pedigree in, 157; suitability in, 158; in regard to speed and stamina, 159; of the mare, 170-186; of the stallion, 187-198; of hunting sires and dams, 269; for getting half-bred stock, 337
Separation of dam and foal, 248
Service of mares, 230; time for, 234
Shape and size of brood-mares, 180
Sheep and cattle, English breeds the best, 97; improvements in our breed of, 100; feeding off the paddocks, 369
Shetland pony, the, 353
Shire-horse, the, *versus* the half-bred, 290-311; comparison of work done by, 324
'Shock,' a, among lambs, 399
Short courses a fallacious test of quality, 81; and long courses, 225
Siderolite, stock of, 267
Sigmaphone and bad temper, 121
Sir Teddy, extraordinary feat of, 352
Sire and dam, their respective influence on the offspring, 200; age of, 202
Site for stud farm, 102
Situation of stud farm, 304
Size of the best stallions, 191; causes of deterioration in, 353
Sleeping-room in the loft, 387
Slipping and sterility, 106-118; a disease and hereditary, 107; examples in proof of my contention, 109; barrenness a consequence of, 111; cases in point, 113; causes of, 114; annual loss from, 116
Small kinds of horses, 350, etc.; distinguished by greater hardihood, 354
Smith, Mr. T. Ashton, a veteran of the hunting-field, 15; anecdote of, 273
Smithfield horse-market in the olden time, 43
Soil of the stud farm pastures, 365, 366
Some good-looking failures, 174
Spanish strain, the, 54; ass, the, 356; great size of, 361

29—2

Speed and stamina, 159; examples of both, *ib.*; increase of in farm-horse desirable, 289; of cart-horse and half-bred on the road, 304; of coach-horse, 313; of omnibus-horse, 315; of post-horse, *ib.*; of van-horse, *ib.*; of various draught-horses compared, 324; of pony, 352; of ass, 357

Speedy mares the best to breed from, 185

Splint, inherited and accidental, 133

Spread of English horses abroad, 91

Stables, the, construction of, 379, 381, etc.; for the stallion, 388

Stable-keeper, employment of by farmers, 287

Stallion, the, condemned at Newcastle, 22; selection of, 187-198; health of, 188; breeding of, 189; pedigree of, 190; size of, *ib.*; failures of large, 192; shape of, 193; effect of inferior, 195; instances of influence on offspring, 201; large not good to breed from, 204; the thoroughbred, 213-226; evils of breeding from your own, 215; instances of the same, 216-219; old and young, good stock got by, 219; age for the stud, 220; treatment of, 253 262; proper boxes for, 388; size of boxes for, 390; yard attached to boxes for, *ib.*; too many kept, 412; more than double what are required, 414; paucity of really good stallions, *ib.*; even those that get winners faulty, 416; evils propagated, 417; defects in stallions at the stud, 421; foreign sales of, 423; my plan to improve, 425, etc.; stallions at Alvediston, 430; covering in 1883-4, 429

Stamina and speed, 159; examples of both, *ib.*

Stephenson, George, on speed, 305

Sterility and slipping, 106-118; soon discoverable in a mare, 108; examples in proof of hereditary tendency to, 110; a consequence of slipping, 111; cases in point, 113; causes of, 114

Stinting mares after service, 233

Stockwell, a great sire, 62; unique instance of a large and yet good sire, 192; his sire and dam, 204; his pedigree and descendants, 215

'Stonehenge,' his book not up to date, 7; on unsoundness, 151; on lame stallions, 256; on training, a capital error of, 393; on the brain of animals, 403; on the blood and blood-vessels, 405 [214

Strains, thoroughbred, past and present, 'Stud-Book, The,' institution of, 55

Stud farm, the: the land, 364-376; the buildings, 377-391

Stud grooms, negligence of, 258

Suckling foal, the, 246

Suffolk Punch, the, 340

Suitability of climate necessary, 95; of sires and dams to each other, 158

Summary of principles of selection, 168

Superiority of English horses over Arabs, 74

Supposed degeneracy of the modern horse an error, 80

Syrian ass, the, 356

Surplice, temper of, 122

Sweating, 393

Swiftness of ass and onager, 357

Sykes, Sir Tatton, his principle of selection, 154; his horses by no means bad, 155; his experiments in breeding, 165

Table, comparative, of speed and work of draught-horses, 324

Tank for rain-water, provision of a, 381

Temper, 119-127; traced through generations, 120; examples of, 121-127; influence of breaking on, 125; good-tempered mares should alone be bred from, 127

Tendency of animals to revert to former types, 138 [183

Terrier, instance of curious action in a,

Time to have mares covered, 228; for foaling, indications of, 244; waste of on the farm, 300, etc.

The Bard, sold to France, high price of, 423

Thoroughbred, the, definition of what constitutes, 199; mare, 199-212; stallion, 213-226; the rate walked by, 282; in the plough, 286; statistics of, 413; mares that should be excluded from 'The Stud-Book,' 422; my plan to improve our breed of, 425; sires and stallions of 1883-4, list of, 439

INDEX.

Top Gallant, Mr. Woodruff's account of, 344
Touchstone, a great sire, 62; and Eclipse, a comparison, 87; at the stud, 166; pedigree and descendants of, 215
Trainers' systems, 224
Transmission of peculiarities, 151
Treatment of mare, foal, and yearling, 238-252; of stallion, 253-262
Trees, kinds of to form shelter plantations about the paddocks, 371; planting out of, ib.; kinds of to be excluded, 372; within the paddocks, 376
Trelawny, Mr. C., a veteran of the hunting-field, 16
Troop-horse and hunter, the, 263-274
Trotting horse, the American, 343, etc.; how reared, 347; and draught-horse, 336-349
Turks, introduction of, 46
Twenty-four mares, an estimate for farmers, 279, 284, etc.
Two-year-olds, riding of, 271

Udders, swollen, how to treat, 249
Unsoundness, 150
Upton, Captain R. D., his 'Newmarket and Arabia,' 64, etc.

Value and cost of horses, 277; of breeding half-breeds, 279
Van-horses, speed of and loads drawn by, 315, 324
Variety of food, 248
Vedette, non-success in his own stud, 216
Venison, lameness of, 256
Ventilation of the stables, 385
Vittoria, temper of, 123
Voltigeur, a great sire, 62; pedigree and descendants of, 215

Wages and work of agriculturists, 292
Waggon-horses, relative speed of and loads drawn by, 324
Wales, H.R.H. the Prince of, in the hunting-field, 18; as breeder and owner, 49
Want of exercise a cause of mortality, 258

War-horse, the ancient, 263
Waste of time on the farm, 300, etc.
Weaning, the time of, 248
Weatherby, Messrs., and 'The Stud-Book,' 54
Weight of cattle, the, 101
Well-bred horses in New Zealand, 94
Whitethorn and blackthorn hedges, 370
White Turk, Place's, importation of the, 46
Whitwick, Mr., and Grace Darling, 177
Why few mares breed good horses, 160
Wild animals in Great Britain, 36
Wild ass, the, 355, etc.
Wild cattle at Chillingham, 98
William IV. and his trainer, 49
Wintonian, trouble to get him to serve, 262
Woodruff, Mr. Hiram, on age, 167; on the American trotter, 343, etc.
Working hours of farm horses and men, 291
Working of pregnant mares not prejudicial, 276
Workhouse, the, and the agricultural labourer, 330
Wreford, Mr., of Gratton, his experiences, 173; failure in breeding from his own stallion, 218
Wrong way to treat stallions, the, 254

Xenophon's work on the horse, 51

Yankee yarns, 344, 345
Yard, the stable, 385; for the stallion, 390
Yearlings, treatment of, 250-252; over-feeding and under-exercising of, 411; all moderate colts should be gelded, 416; list of, at Alvediston, 436
Yorkshire horses, 368
Youatt on selection, 140
Young Melbourne, lameness of, 256
Young mares, treatment of, 240; stallions, good stock got by, 219
Young Trumpeter, temper of, 120

Zebra crossed with mare, a, 197

THE END.

BILLING AND SONS, PRINTERS, GUILDFORD.

S. & H.

www.ingramcontent.com/pod-product-compliance
Lightning Source LLC
Chambersburg PA
CBHW022057300426
44117CB00007B/494